Science 7
Directions

Science Directions 7
Science Directions 8
Science Directions 9

Science Directions 7 Teacher Resource Package
Science Directions 8 Teacher Resource Package
Science Directions 9 Teacher Resource Package

John Wiley & Sons Canada Limited/Arnold Publishing Ltd.

SI Metric

Our Cover Photograph

Does the insect on our cover look familiar to you? It should because it is a type of insect that we have all encountered—a mosquito.

The head of the mosquito has been magnified thousands of times by a scanning electron microscope (SEM). This kind of microscope magnifies objects much more than a compound light microscope such as the one you use in your junior high school science classes. Instead of light and a glass lens to bring the object into focus, the electron microscope uses a beam of electrons and magnets. The microscope does not produce images in colour, so the colour is added later.

The yellow-fever mosquito, *Aedes aegypti*, lives close to humans and lays its eggs in water that collects in discarded rubbish such as tins, bottles, or tires.

The male yellow-fever mosquito, such as the one shown here, feeds on fruit and nectar. The female yellow-fever mosquito needs meals of blood in order to reproduce. She can transmit the yellow-fever virus, which these mosquitoes often carry, while feeding on human blood. (Fortunately, this kind of mosquito is not found in Canada.)

Science Directions 7

PROGRAM CONSULTANT

Douglas A. Roberts
Faculty of Education, University of Calgary

AUTHOR TEAM

Wilson C. Durward • Eric S. Grace
Gene Krupa • Mary Krupa
Alan J. Hirsch • David A. E. Spalding
Bradley J. Baker • Sandy M. Wohl

CONTRIBUTING AUTHOR

Jean Bullard

John Wiley & Sons
Toronto   New York   Chichester
Brisbane   Singapore

Arnold Publishing
Edmonton

Canadian Cataloguing in Publication Data
Main entry under title:

Science directions 7

ISBN 0-471-79577-1

1. Science—Juvenile literature. 2. Science—
Problems, exercises, etc.—Juvenile literature.
I. Durward, Wilson C., 1944-

Q161.2.S23   1989   j500   C89-093856-3

Science Directions 7 Project Team
Project Director: Trudy Rising
Project Manager: Grace Deutsch
Project Editor: Tilly Crawley
Associate Editor: Wilson Durward
Developmental Editors: Jane McNulty, Mary Kay Winter, Bruce Bartlett
Bias Review: Jane McNulty
Copy Editors: Elizabeth Reid, Freya Godard
Word Processing: Diane Klim, Sharon Oosthoek
Design: Julian Cleva
Production Co-ordinator/Art Director: Francine Geraci
Proofreader: Susan Marshall
Picture Research: Jane Affleck
Typesetting: Lithocomp
Assemby: Blue Line Productions
Film: Graphitech
Manufacturing: Arcata Graphics
Field Test Preparation Co-ordinators: Ann Adair, Shona Wehm
Workshop Co-ordinator: Janet Mayfield
Field Test Design and Preparation: Bruce Campbell, Arnold Publishing
Ltd.; Presentation Plus Desktop Publishing Inc.

Printed and bound in the United States of America.

2 3 4 5   AG   98 97 96 95

Contents

Acknowledgement

Science Directions 7 was developed specifically for the Alberta Junior High Science Program. The book is the result of the efforts of a great many people.

First, we would like to thank Alberta Education for reviewing this project at its various stages of development. We thank particularly the Junior High Science Advisory Committee, and Mr. Bernie Galbraith, Program Manager, for their thorough analysis of the first draft, the field test (pilot) material, and the final pages.

The efforts of two other groups were invaluable in developing the book. The Teacher Resource Package authors reviewed each Unit and made many helpful suggestions throughout manuscript development. They are: Monika Amies, Calgary; Stephen Jeans, Calgary; Laurier Nobert, Rocky Mountain House; Paul Liska, Lethbridge; Jeannette Kucher, Edmonton; and David Turner, Edmonton.

Each Unit of **Science Directions 7** has been field tested. We would very much like to thank the following teachers and their students for using our materials in draft form and giving us their reactions to assist us in the development of the final text: Paul Dvorack, St. Margaret School, Calgary (Units One and Three); Herb Bergerman, St. Matthew School, Calgary (Units One and Five); Nancy Hope, Crystal Park School, Grande Prairie (Units One and Five); John Woycenko, St. James School, Edmonton (Units One and Five); Mark Collard, Harold Panabaker Junior High School, Calgary (Units One and Six); Len Kachuk, Blessed Sacrament School, Wainwright (Units Two and Three); Kay Jauch, Allendale School, Edmonton (Units Two and Five); Sister Elaine Cole, St. Basil Junior High School, Edmonton (Units Two and Six); Craig Herbert, Winterburn School, Winterburn (Units Three and Four); David Turner, McKernan Junior High School, Edmonton (Units Four and Five); Elizabeth Arthur, Gilbert Patterson Junior High School, Lethbridge (Units Four and Six); Donna Cruden, Norm Howes, and Larry Hartel, Glendale School, Red Deer (all six Units); Dale Makar, Bruce McLeod, and John McLean, Annie Gale Junior High School, Calgary (all six Units).

We also thank the following reviewers for assisting us in ensuring that the content was pedagogically sound and scientifically accurate: Dr. Nancy Flood (Unit One); Dennis Person of the Northern Alberta Institute of Technology, Jean Bullard, and Pamela Kay (Units Two and Three); Jean Bullard (Unit Four); Dr. Fred H. Wolfe and Dr. Michael E. Stiles of the University of Alberta (Unit Five); and Dixon Edwards of the Alberta Research Council and the Alberta Geological Survey; Walter G. Kemball (Unit Six); and David Gray (all Units).

The Special Features and Career Features were prepared by Barbara Hehner, Lesley Grant, and Julie E. Czerneda. Their research was assisted by Dr. Ron Weir, Consultant, Canadian Wildlife Service; Jacqueline Starink; Eric Kiisel, Atomic Energy of Canada; André Simard, Acrobatic Director, National Circus School, Montreal; Marc Garand, Scarborough Olympians Gymnastic Club; Debra Mathews, Canadian Broadcasting Corporation; Anne Bondarenko, Cirque du soleil; Sherry Draisey, Spar Aerospace Limited; Hanna Pilar, Association of Professional Engineers of Ontario; Vicki Keith, marathon swimmer; Dr. John Smith, Hospital for Sick Children, Toronto; Daniel Crichton, Sheridan College; Professor Doug Cunningham, University of Guelph; John B. Webster, architect, Edmonton; and Robert Babiarz, forest resource technologist. We thank them all for their contribution.

We thank Meru Brunn for permission to use the poem on page 271, and Gerry Sieben for the use of Activity 1-10, The Living Submarine. Andrea Carter is thanked for the lovely illustrations she produced for our program proposal. Lastly we especially thank Christine Suchy-Shantz for her innovative ideas for the cartoon strips you see used throughout.

The Authors

This is a very exciting time to be teaching science. It is also a very demanding time. Science educators all over the world are being challenged to rethink the goals, purposes, and processes of science teaching. The importance of scientific literacy for all citizens has become increasingly clear. As well, the challenge to demonstrate the relationships among science, technology, and society is generally accepted as an essential in the teaching of science.

Science Directions responds to these new challenges. This program gets students involved in the processes of science, technological problem-solving, and in discussions relating science and technology to social issues. I have found it very impressive to watch teachers working with these materials. I have seen students develop and defend different value positions in debates. And I have seen both girls and boys equally involved in the Activities.

Science Directions provides a balanced approach to science by emphasizing three important goals of science education in each book. Some Units concentrate on the nature of science and science processes. Others place their main emphasis on the relationship between science and technology. And still others expose students to science-technology-society (STS) understandings.

It has been a pleasure for me to work with these creative, solid, and very teachable materials. With texts of this kind, science teachers can provide the kind of balanced program that meets the challenges and demands of science education for the 1990s and beyond.

Douglas A. Roberts
Program Consultant

Features of Science Directions 7

- **Science Directions 7** has six Units. Each introduces a major area of science, using an appropriate balance in emphases: nature of science, science and technology, and science-technology-society (STS).
- Many and varied *Activities* are included in each Unit: formal investigations, informal discoveries, technical challenges, among others. Some Activities invite the students to solve practical problems; others provide opportunities for them to design their own experiments.
- The Activities are followed by three levels of questioning:
  Finding Out leads students to consolidate their results and observations;
  Finding Out More challenges them to reach logical conclusions and to reason beyond the immediate and obvious results they have obtained;
  Extension encourages investigations or explorations in new directions.
- Within the text, *Probing* questions help enhance the students' understanding of textual and/or visual material by leading them to think through related problems.
- The *Did You Know?* heading introduces brief statements of scientific, technological, or societal interest related to the concepts under discussion.
- After every two Topics within a Unit a *Checkpoint* of graded questions provides review and reinforcement of the ideas discussed.
- Each Unit ends with a *Focus* to review the subject matter in point form, a *Backtrack* to test the students' grasp of the content and process skills, and a *Synthesizer* to challenge the students to think through and beyond the knowledge, skills, and processes developed in the Unit.
- Within each Unit is a two-page special feature entitled *Science and Technology in Society*. While the text at all times attempts to use concrete examples to explain abstract concepts, the special features emphasize actual situations and highlight real-life applications of the scientific ideas under consideration in the Unit. This reinforces the students' understanding that the concepts they are investigating have an immediate impact on their society and themselves.

- Each Unit also contains a career feature, entitled *Working with...or Working by...*, that will stimulate students to consider how they might develop their own interests and capabilities into rewarding careers in science.
- Throughout the text, new terms are given in **bold face** and defined in context; they also appear in the *Glossary* and in the *Index*.
- Appendices, entitled *Skillbuilders*, on units in SI, science process skills, measurement, graphing, and use of the microscope, provide useful support and reference for the students.
- The text opens with extensive directions on the safety rules to be observed at all times in science classes. The students should become thoroughly acquainted with these safety rules, and the reasons for them, at the start of the school year. In addition, a **CAUTION** is included when special care must be taken in using equipment or materials.
- The *Teacher Resource Package* provides guidance on the emphases of the program and offers hints on various teaching strategies. The package is designed to provide a wealth of useful information for both experienced science teachers and those new to the program.

Safety in the Science Classroom

Your school laboratory is designed so that you can perform science experiments in safety—provided you follow proper procedures. Just as you must be careful in your kitchen at home, so too you must be careful as you handle materials and use the equipment in the laboratory.

The Activities in this textbook have been tested and are safe, as long as they are done with proper care. When special attention is needed you will see the word **CAUTION** with a note about the particular care this Activity requires.

Your teacher will give you specific information about the safety routines used in your school and will make sure that you know where all the safety equipment is.

If you follow these guidelines and general safety rules, along with your own school's rules, you will have an accident-free environment in which to enjoy science.

General Safety Precautions

- Work quietly and carefully; accidents, as well as poor results, can be caused by carelessness. Never work alone; if you have an accident, there will be no one there to help you.
- Tie back loose hair, roll back and secure loose sleeves, and make sure you are wearing shoes that cover your feet as much as possible. Don't wear scarves or ties, baggy clothing, earphones, or jewellery in the laboratory—these can catch on equipment and knock it down.
- Try not to wear contact lenses during experiments involving chemicals.
- Inform your teacher of any allergies, medical conditions, or other physical problems you may have.
- Never eat or drink in the laboratory.
- Do not do laboratory experiments at home unless directed to do so. Some students have been seriously injured when performing experiments on their own.

Safety equipment

- Listen carefully to your teacher's instructions on when and how to use the safety equipment: safety glasses, protective aprons, fire extinguishers, fire blankets, eye-wash fountain, and showers.

- Make sure you know where the nearest fire alarm is.

Before you begin an Activity
- Read through the entire Activity so that you know what to do.
- Make sure the work area is clean before you start. Clear everything (books, papers, personal belongings) except your textbook and notebook from the work area.
- Do not begin an Activity until you are instructed to do so.

Handling materials
- Touch substances only when told to do so. What looks harmless may, in fact, be dangerous.

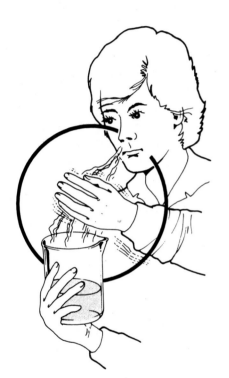

- When you are instructed to smell a substance in the laboratory, follow the procedure shown here. Only this technique should be used to smell substances. Never place the substance close to your nose.
- Never pour liquids while holding the containers close to your face. Place a test tube in a rack before pouring liquids into it.
- If any part of your body comes in contact with any harmful chemical or specimen, rinse the area immediately and thoroughly with water. If your eyes are affected, do not touch them but rinse them with water immediately and continuously for at least 10 min.
- Wash your hands after you handle substances.
- Clean up any spilled substances immediately, as instructed by your teacher.
- Never pour harmful substances into the sink. Follow your teacher's instructions about how to dispose of them.

Handling a heat source
- Whenever possible use electric hot plates with thermostatic controls.
- To heat a test tube using a hot plate, place it in a beaker of water on the hot plate.
- Heat material in heat-resistant glass containers only. Make certain the glass you use for heating is Pyrex or Kimax and is not cracked.
- Always keep the open end of a test tube pointed away from other people and from yourself.
- Never allow a container to boil dry.
- Pick up hot objects carefully using tongs.
- Make sure that the hot plate is turned off when not in use.

- Always unplug electric cords by pulling on the plug, not the cord. Report any frayed cords to your teacher.
- Be careful how the cord from the outlet to the hot plate is placed. Make sure it is not in a place where anyone could trip over it.
- If you burn yourself, apply cold water or ice immediately.

Bunsen burners need special care
- If Bunsen burners are used in your science classroom, use them only when instructed to do so. Observe the precautions that follow at all times.
- Obtain instructions from your teacher on the proper method of lighting and using the Bunsen burner, and make sure you understand the instructions.
- Check that there are no flammable substances in the room before you light the burner.
- Use tongs to hold a test tube that is being heated in the flame. Point the test tube away from yourself and other people, and move the test tube back and forth over the flame so that heat is distributed evenly. Be ready to turn the Bunsen burner off immediately if necessary.
- Never leave a lighted Bunsen burner unattended.
- Never heat a flammable substance over a Bunsen burner.
- Follow your school's rules in case of fire.

Other equipment
- Your teacher will provide stoppers that are already fitted with thermometers and tubing for use in the Activities in Unit Four. But remember never to force glass tubing or a thermometer into a dry stopper. The hole in the stopper must always be lubricated with soapy water or glycerine.
- Never use cracked or broken glassware. If glass does break when you are using it, follow your teacher's instructions in clearing it away.
- Watch for sharp or jagged edges on equipment.
- After each experiment, clean all equipment and put it away. Do not leave equipment that is not in use lying around the work area.
- Report to your teacher all accidents (no matter how minor), broken equipment, damaged or defective facilities, and suspicious-looking chemicals. In this way you will be taking responsibility not only for your own safety, but also for the safety of those who use the laboratory after you.

Have you ever been amazed to see that newborn animals such as this calf are able to stand up and even move about within minutes of their birth? In Unit One you will find out about the processes that keep living things— including you—alive. And you will discover what makes living things different from non-living things.

Some of the living things you will study in Unit One build quite elaborate structures to shelter themselves and their young. People build structures, too. In Unit Two you will think about the structures that people build, and compare them with structures in the natural world. We need structures for shelter and protection, but we build structures for many other reasons as well. Discover how and why architects, building contractors, designers, inventors, and aerospace engineers use different shapes and materials.

A beautiful place-setting— but how did the magician pull the tablecloth out from under the plates and the cup without breaking anything?

The satellite continually circles the Earth. Why doesn't it fall back to Earth?

What do the magician and designers of the satellite have in common? They both make use of their knowledge of forces, but for very different purposes. In Unit Three you will explore how an understanding of forces and the motion they cause can put people and satellites into space, and provide the magician with a way to entertain you.

Why are things hot and cold? In Unit Four you'll find out. You'll also do some inventing.

Inventing and using precision instruments is a part of science and modern technology. Being able to measure accurately has given us many of the useful appliances we have today. You'll investigate one of these instruments—the thermometer. You'll invent and use a thermometer of your own design.

Some living things are so small that humans can't see them with the unaided eye. Some of these microscopic living things help improve our lives. Yet others cause serious illnesses.

All life as we know it occurs on, just under, or just above the surface of planet Earth. We think we know Earth's surface—after all, we live on it. Yet we seldom realize that it is changing constantly. What are some of the things that cause such changes? Discover them in Unit Six.

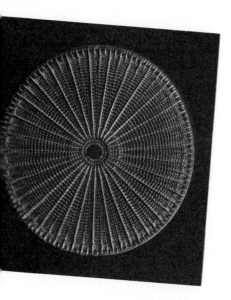

Many of these tiny life forms affect the food we eat. Find out about microscopic living things and how they affect your life and your food in Unit Five.

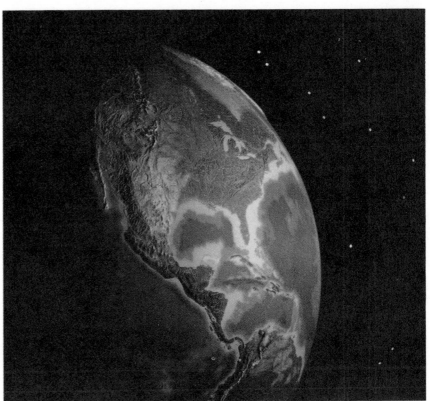

Science Directions 7 lets you explore in several different directions. In grades 8 and 9 you'll take other paths in your science explorations.

Our world is both beautiful and interesting. We hope that by exploring it in **Science Directions** you will understand it better and share in helping to keep it truly beautiful.

The Authors

A Close Look at Life

How can you tell if something is alive? What clues might help you to decide? Think about it. Just what do you know about living things? In this Unit you will be investigating the features of living things.

Look closely at the objects in the photographs on these two pages. Keeping your "clues for life" in mind, suggest which of the photographs show living things.

(a)

(d)

(b)

(c)

(i)

(j)

(e)

(h)

(f)

(g)

Characteristics of Living Things

Did you know that *all* of the photos on the previous pages contained living things? Yes, even (c), (f), (i), and (j). Below are some close-up shots of the same objects in (c), (f), (i), and (j).

This Topic is an overview of the main characteristics of living things. Later in this Unit you will be finding out about some of these characteristics in more detail. But first, test yourself—what do you already know about living things?

What can you observe in these photographs to tell you that photographs (c), (f), (i), and (j) were living things?

(c)

(f)

(j)

(i)

Look on page 70 to find out what these living things are.

Thinking about Living Things

1. Make a class list of as many different kinds of living things as you can think of in just thirty seconds.
2. Now list as many features as you can that they all have in common.
3. On the chalkboard, combine the lists of features, or **characteristics**, you and your classmates have made into one list. Copy the list into your notebook.

Finding Out

You and your classmates probably thought of the most important characteristics that living things have in common. Compare your list with the characteristics described on the next few pages to see if you missed any. (You may have thought of characteristics in addition to the ones listed here.)

What Do Living Things Have in Common?

Before you begin your study of living things, you need to become familiar with two scientific words that are used frequently when talking about living things. The first of these words is **organism**, which is simply another, scientific name for a living thing. The second word is **species**. Every organism is a member of a species. You are a member of the species human being. All domestic cats are members of the same species. Organisms that are members of the same species have many characteristics in common.

Now read the overview of the characteristics of organisms on the next few pages.

All organisms grow. You know that you have grown from a baby, and you can tell that you are still growing when your clothes no longer fit! Every living thing grows to be approximately the same size and shape as its parents. Or, to put it another way, every member of a species grows to be similar in size and shape to other members of its species.

Members of a species do not look exactly like each other. If they did, you would not be able to tell human beings apart. But members of a species are recognizably like each other.

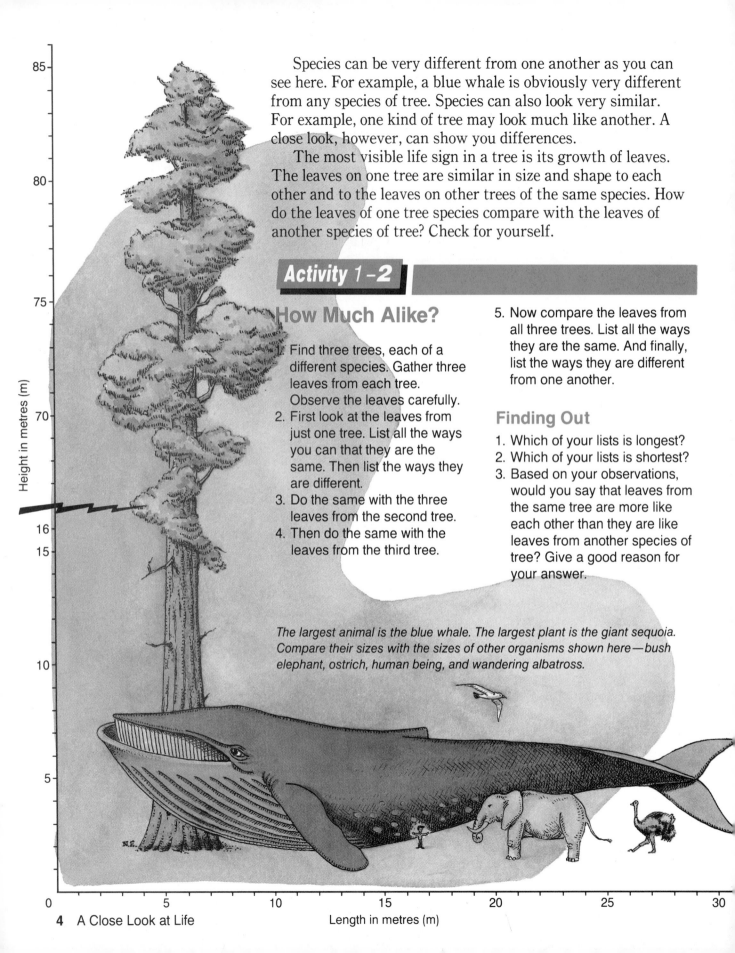

Species can be very different from one another as you can see here. For example, a blue whale is obviously very different from any species of tree. Species can also look very similar. For example, one kind of tree may look much like another. A close look, however, can show you differences.

The most visible life sign in a tree is its growth of leaves. The leaves on one tree are similar in size and shape to each other and to the leaves on other trees of the same species. How do the leaves of one tree species compare with the leaves of another species of tree? Check for yourself.

Activity 1–2

How Much Alike?

1. Find three trees, each of a different species. Gather three leaves from each tree. Observe the leaves carefully.
2. First look at the leaves from just one tree. List all the ways you can that they are the same. Then list the ways they are different.
3. Do the same with the three leaves from the second tree.
4. Then do the same with the leaves from the third tree.
5. Now compare the leaves from all three trees. List all the ways they are the same. And finally, list the ways they are different from one another.

Finding Out

1. Which of your lists is longest?
2. Which of your lists is shortest?
3. Based on your observations, would you say that leaves from the same tree are more like each other than they are like leaves from another species of tree? Give a good reason for your answer.

The largest animal is the blue whale. The largest plant is the giant sequoia. Compare their sizes with the sizes of other organisms shown here — bush elephant, ostrich, human being, and wandering albatross.

Height in metres (m)

85
80
75
70
16
15
10
5

0   5   10   15   20   25   30
Length in metres (m)

All organisms move. In an organism, **movement** is any motion or activity that changes the organism's shape, position, or place. Animals move from place to place. Such movement is obvious. But even slight movements, such as the movements you make in breathing, are evidence that an organism is alive. Plants and tiny organisms that you can see only with a microscope also move. Some move quickly and others slowly, but they all move.

Like other plants, this bean seedling bends slowly towards the sun's rays.

*Using a microscope, you can see how tiny organisms like the paramecium move. The **cilia**, hairlike structures surrounding its body, wave back and forth, moving the paramecium through the water.*

cilia

The cheetah's very flexible backbone allows it to use long strides. A cheetah can move for short distances at speeds up to 110 km/h.

Probing

LOOK! It's alive.

Not really.

Imagine that you are caring for a young child who has just come home from a party. She was given a balloon at the party, and she asks you to blow it up. You do so, tie the balloon off, and gently throw it to the child. It bounces across the floor to her. "Look," she says, "it's alive! It grew big, and now it's moving all over the room!"

Maybe you'll decide not to spoil her fun by telling her that the balloon is not alive. But explain it to yourself: write in your notebook how you would explain that a balloon is not alive even though it grew and moved.

All organisms reproduce. Organisms of every kind produce offspring similar to themselves. Organisms continue to exist by producing offspring of their own kind. Organisms reproduce in different ways.

*Flowers are the reproductive parts of plants. "A" contains the male reproductive part. "B" contains the female part. When "A" and "B" unite, the result is a **seed**. If it has the right conditions for growth, a seed will produce a new plant.*

Some organisms, such as this tiny amoeba, can reproduce by splitting in half.

Most animals, like these male and female Japanese beetles, must mate before they produce offspring.

All organisms produce or take in food. Some organisms are able to produce their own food. They use the sun's energy to make food from carbon dioxide and water. This process is called **photosynthesis**. Organisms that cannot produce their own food obtain it by eating other organisms. All organisms use the food to build body parts, and to provide energy to move, grow, and reproduce.

These prairie crocuses are absorbing energy from the sun to produce the food they need to grow.

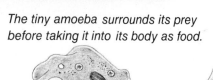

The tiny amoeba surrounds its prey before taking it into its body as food.

All animals need to find and eat food—meat, plants, or both—to provide them with energy.

All organisms respond to stimuli in their environment. Organisms sense things that happen in their surroundings and react to them. For example, in the school environment, a friend calling your name is a **stimulus** that you sense and to which you respond. In an outdoor environment, a mosquito landing on your skin is also a stimulus to which you respond. Think about it—which senses are you using as you detect each of these **stimuli**?

What do you think the response of a mouse in the grass might be to the stimulus of the hawk overhead?

The stimulus of a fly touching the sensitive inner leaf hairs of the Venus flytrap causes the two halves of the leaf to close suddenly, trapping the fly inside.

Activity 1-3

Mimicking Life's Signs

Now that you have spent quite a bit of time thinking about the characteristics of living things, consider the following pairs of situations and answer the Finding Out questions about them.

A. A ball bounces across the road.
   A rabbit hops across the road.

B. A snowball grows larger.
   A young child grows taller.

C. A drop of oil splits in half.
   An amoeba splits in two and forms two new amoebas.

D. A compressed spring rebounds.
   The leaf of a Venus flytrap closes.

Finding Out

1. Write in your notebook
   (a) the name of the living thing in each pair;
   (b) which characteristic of life it is showing in the situation;
   (c) what other characteristics of living things it has.
2. Give a definition of life based on the characteristics of living things.

What Are Living Things Made Of?

You can observe directly all the characteristics of living things you have just been reading about. Two other characteristics of living things are more difficult to observe, but they are just as important.

All organisms are made of cells. If you look at a small part of any living thing through a microscope, you will see that it is made up of many tiny units. These units are called **cells**. All things that are living or *have been living* are made up of cells. Non-living things are not made up of cells. Think about that phrase "have been living." You can now probably distinguish between most things that are living or non-living, but what about things that are "dead"? A leaf that has fallen from a tree—is it non-living or dead? When we say something is dead, we mean that it once had life. If you look at things that were once alive but now are dead, such as the wood in your desk, or a leaf after it has fallen from a tree, you will still be able to see the areas where cells used to be.

A plant cell magnified many times under a microscope: its general shape is rectangular.

(a)

(c)

(b)

One-celled bacteria may be (a) spherical like a ball, (b) rod-shaped, or (c) spiral in shape.

As you can see, a human nerve cell is very irregular in shape.

Probing

What characteristics of living things have dead things lost? What characteristics do they still have?

As different as a worm, a log, an ant, and a leaf are, their chemical make-up is similar.

All organisms have a special chemical make-up.

Worms, ants, sharks, rose bushes, and all other organisms are made up of the same **chemicals**—carbon, nitrogen, hydrogen, and oxygen. Organisms may contain other chemicals as well, but these four are common to all living things. These chemicals are arranged in similar ways in all organisms. Only things that are living or have been living are made of these special chemicals arranged in these special ways.

Activity 1–4

Living, Non-living, or Dead?

Now that you've looked at some of the different ways organisms show the characteristics of life, see how good you are at telling living things from non-living and dead (or no longer living) things. In some cases, it may not be easy. Classify the following objects as living, non-living, or dead:

earthworm   leaf from an oak tree
maple tree   sun
lava rock   salt crystal
blue jay   piece of charcoal
bean seed   apple
banana   strand of hair
candle

Make a table like the one shown here to record your information or **data**. One classification has been done for you, using the balloon example you've already thought about.

Finding Out

Go back to the list of characteristics of living things you made in Activity 1–1. Can you now add to the list? If so, make a new list of the characteristics of living things.

Table 1

OBJECT	LIVING/NON-LIVING/DEAD	REASON FOR MY CLASSIFICATION
balloon	non-living	a balloon cannot reproduce more balloons like itself

A Close Look at Growth and Reproduction

One characteristic of life that we see all around us is growth. Growth means more than just getting bigger. It also means changes—all the changes that an organism goes through during its life. Think about the changes in shape between a tadpole and an adult frog, a fawn and an adult deer, or a young corn plant and a mature one. The tadpole develops legs and air-breathing lungs as it becomes a frog; the male fawn grows antlers as it becomes an adult deer; the corn plant develops corncobs. All these changes are a part of growth.

Growth is part of the **life cycle**. The life cycle of an organism includes the process of reproduction, the pattern of growth, and the adult life of the organism. Each member of a species has a similar life cycle—all tadpoles go through the same stages to become frogs. The life cycles of different species are different—only corn plants can develop corncobs.

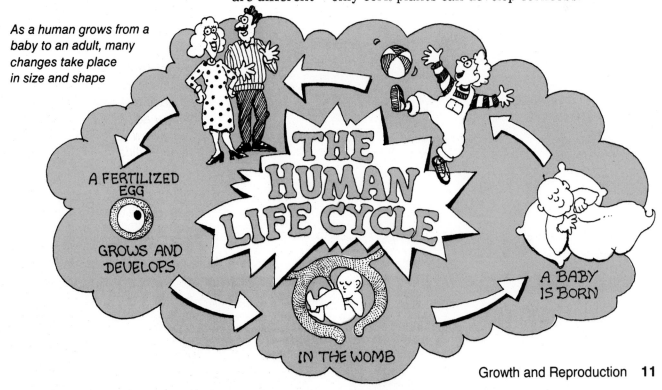

As a human grows from a baby to an adult, many changes take place in size and shape

Patterns of Growth

Have you ever noticed how big a puppy's feet and ears are for its body size? Some people try to determine how large a puppy will grow by looking at the size of its feet. If you have a pup with big feet, you may well soon have a very large dog to feed!

What about humans? No, we do not look at babies' feet to see how tall they are going to become. But it is true that the body proportions of human infants and those of adults are also quite different. This is because the parts of human bodies grow at different rates.

Probing

Look closely at the diagram of the infant and the adult and think about these questions.
1. Which parts of the body grow the most between infancy and adulthood?
2. Which parts of the body grow the least?
3. Explain why a baby's head is so big for its body size compared with an adult's.

If you were to draw a baby standing upright and at the same height as an adult, it would look like the baby in the illustration. The baby looks strange drawn like this, but it does allow you to make some interesting observations!

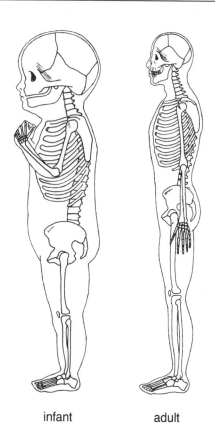

infant   adult

Variation

Members of a species are never exactly like one another. There is a great deal of **variation** within each species. Think of the variations in size, colour, and markings in dogs. In the next Activity you'll be able to examine a single source of variation in humans —the human handspan.

How Wide Is Your Handspan?

Problem

How much variation in the size of handspan is there in the class?

Materials

a ruler
graph paper
pencil
a sheet of plain paper

Procedure

1. Spread out your hand on a piece of paper as shown. Mark a point at the tip of your thumb and your little finger. (Do not include the length of your fingernails.)
2. Join the two points, and measure the line length in millimetres. This length is your handspan.

3. Record the individual handspans on the chalkboard in order of size. Arrange them in a table by 5-mm groups. Then record the number of handspans in each group.

Table 2

HANDSPAN IN MILLIMETRES (mm)	NUMBER OF STUDENTS
160–164	
165–169	
170–174	
175–179	

4. Using graph paper, draw a bar graph of the data in your table. Look at *Skillbuilder Four* if you need to review how to make a bar graph.

Finding Out

1. Which group of handspan sizes had the largest number of students?
2. Which group of handspan sizes had the smallest number of students?
3. What was the difference, in millimetres, between the handspan sizes of these two groups?

Finding Out More

4. For most human beings, handspan size is probably not important one way or the other.
(a) Suggest a way of life that would make having very small hands an advantage.
(b) Suggest a way of life that would make having large hands an advantage.
5. List other variations in human hands. Consider shape, length of little finger, and so on.

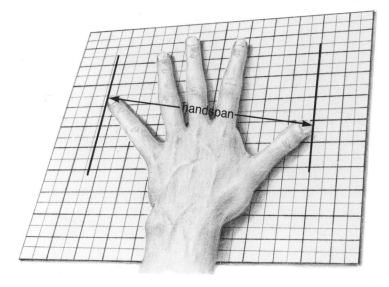

How to measure your handspan.

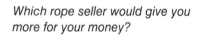

Which rope seller would give you more for your money?

Rates of Growth

It's hard to imagine, but the students in this photograph are all the same age. These students are at the growth stage of the human life cycle, when the greatest variation in growth occurs. Growth in human beings involves an increase in both height and body mass. **Mass** is the amount of matter or material in an object. In humans, height is measured in centimetres; mass is measured in kilograms.

Teenagers often experience a rapid increase in height and body mass at the same time. This "growth spurt" happens between the ages of 11 and 13 in many girls and between 14 and 16 in many boys. However, it is quite normal for the growth spurt to occur much earlier or much later.

In one school, all grade 10 students were asked to bring information on their own growth. In the following Activity you will analyse the information Peter Boyko brought on his increase in height from birth to his present age. If you would like to find out about human growth by studying increase in body mass, do Skillbuilding Practice 4-5 on page 366.

Measuring Growth

Peter's parents kept a record of his growth on a wall chart. The table was made from this record. The line graph shows the same information about Peter's growth as the table does, but in a way that makes it easier to see a pattern in the way Peter grew.

The line along the bottom of the graph—the **horizontal axis**—shows age. The line up the left-hand side—the **vertical axis**—shows height. A diagonal line on a graph would indicate that there is a close relationship between the two things being shown on the graph. On this graph, a straight diagonal line would have shown that Peter grew the same amount every year.

Finding Out

1. What does the shape of the line on the graph show you about Peter's growth pattern?
2. When was Peter's "growth spurt"?
3. How many centimetres did Peter grow in the year of his growth spurt?
4. Did Peter's growth spurt occur at the same time as that of most boys? Or was it earlier or later?
5. When was Peter's period of slowest growth? Think of a reason why this might be. (Hint: Use the graph and what you know about growth patterns to help you answer this question.)

Table 3 *Peter's age and height*

AGE IN YEARS	HEIGHT IN CENTIMETRES (cm)
0	60
1	75
2	85
3	94
4	100
5	105
6	108
7	112
8	118
9	125
10	130
11	140
12	145
13	171
14	173
15	174
16	175

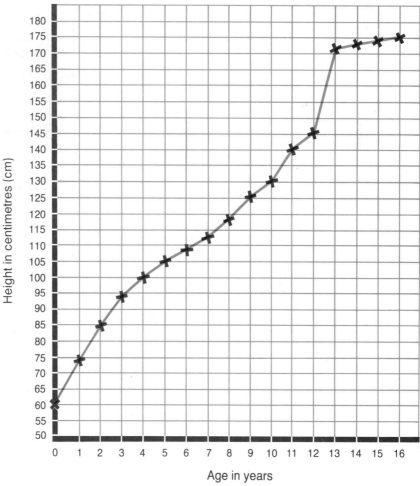

Peter's growth pattern

Growing in Stops and Starts

Although the rate of Peter's growth was much greater in some years than in others, it was a fairly smooth process. For some other organisms, growing is more a matter of stops and starts.

Animals such as lobsters, crayfish, crabs, and insects have a skeleton *surrounding* their bodies instead of inside it, as we do. This hard, outer shell does not grow, and it prevents the growth of the animal within. How, then, does such an animal grow? It sheds its skeleton and grows a new, larger one. Therefore it grows in a series of spurts, between the shedding of the old skeleton and before the hardening of the newly produced skeleton.

The growth of the water boatman

Erica Bartlett has always been interested in insects. Even before she went to school, she spent a lot of time making mazes for ants, and watching the insects around the pond of the farm where she lives. For a grade 7 science project, she measured the growth of a small insect, called the water boatman, that lived in the pond.

Probing

1. What does the shape of the line on the graph tell you about the pattern of the water boatman's growth?
2. Why does its growth follow this pattern?

How Living Things Grow

The illustration shows how a single cell reproduces. 1. is the cell before it starts to reproduce. 2, 3, and 4 show some of the steps the cell passes through as it splits into two. 5 shows that at the end of the process there are two similar cells.

A famous scientist, George Simpson, once described the difference in the way living and non-living things grow. He contrasted the pattern of growth in organisms with the building of a brick wall.

"The process of growth in living things is not like building a wall of bricks. In the growth of organisms, as soon as a 'brick' is added, the brick itself adds others around it."

Simpson is describing a very important process. Each of the cells that make up an organism can reproduce more cells like itself. It is this multiplication of cells that makes growth possible.

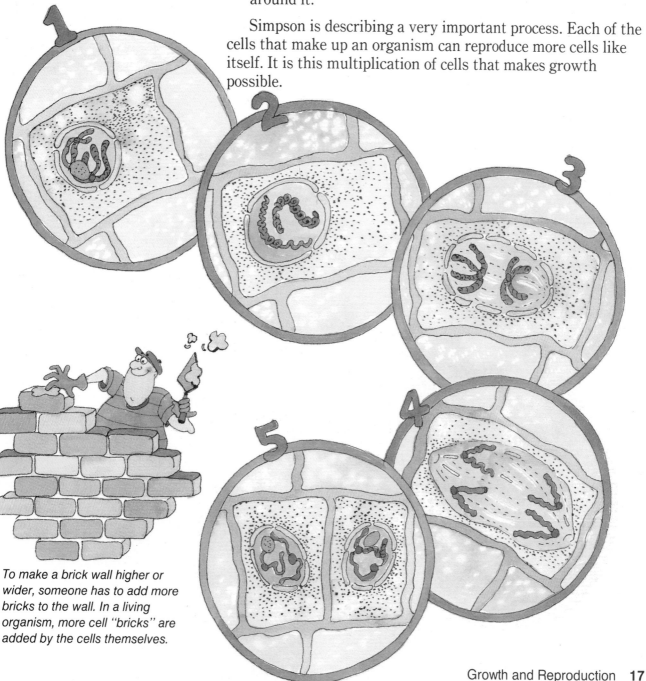

To make a brick wall higher or wider, someone has to add more bricks to the wall. In a living organism, more cell "bricks" are added by the cells themselves.

Growth for Repair

A kind of growth that is important to all organisms is the repair of injured body parts. If you have ever cut yourself, sprained your wrist, or broken a leg, you have experienced this process. Skin, muscle, and bone can repair themselves by growing new cells.

Some animals even have the ability to grow whole new parts. This ability is called **regeneration**. Following an injury, almost any piece of the small flatworm, planaria, can grow into a whole new worm. Many other **invertebrate** animals (animals without backbones) can also grow new parts. For example, an earthworm can regenerate (grow) a new tail section if it is cut at any point after the first 40 segments of its body. In some starfish, a single arm can grow into a whole starfish. Crayfish, crabs, and lobsters can regenerate missing claws, antennae, or legs.

The ability of most **vertebrates** (animals with backbones) to regenerate parts is limited. Most vertebrates cannot grow whole new limbs. Salamanders, however, are an exception. They can grow replacement limbs and tails. Some lizards can also regenerate a tail. Such organisms can lose a tail without permanent damage.

Probing

The growth on this plant is called a gall. Find out what could have caused this growth. Is it unhealthy for the plant?

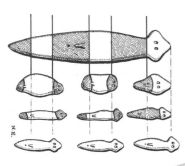

Another planaria can grow from any piece of the injured animal's body.

How does losing its tail help the salamander to survive?

Other Growth Patterns

People and many other species have similar features when they are young, as they are growing, and when they are adults.

Like humans, horses look quite similar when they are young and when they are adult or "mature."

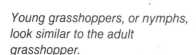

Young grasshoppers, or nymphs, look similar to the adult grasshopper.

But in other species the adult can look entirely different from the young. Look at the illustrations on the next page showing the dramatic changes that occur during the life cycles of the Colorado beetle and the leopard frog.

The type of growth pattern each species has is important for its survival. If you look closely at the world around you, you will see an amazing number of organisms with life cycles similar to the ones shown here and on the next page.

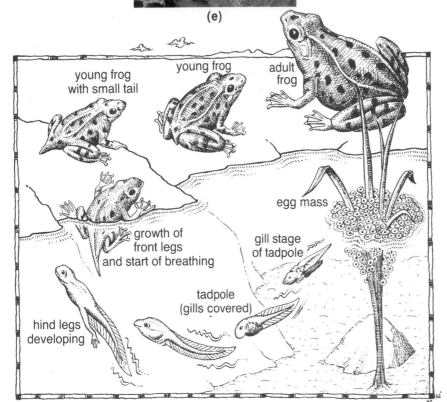

The Colorado potato beetle develops in four different stages—egg, larva, pupa, and adult.
(a) eggs on a potato leaf
(b) eggs and a newly hatched larva
(c) larva in the soil going to the pupa stage
(d) the pupa stage
(e) the adult beetle

Stages in the life cycle of the leopard frog. How many changes can you identify as the frog develops?

young frog with small tail
young frog
adult frog
growth of front legs and start of breathing
egg mass
gill stage of tadpole
hind legs developing
tadpole (gills covered)

Activity 1-7

Growing Greens

Problem

All animals need food, water, warmth, and many also need parental care in order to grow. Plants, too, need a favourable environment. What conditions are necessary to start a seed growing?

Materials

3 small beakers
marking pencil
paper towelling
6 seeds: radish seeds or bean seeds
a small pie plate or shallow tray
a box big enough to cover one beaker

Procedure

1. First, make a list of the conditions you think are necessary for the seeds to begin to grow. Write your list in your notebook.
2. Soak four seeds overnight in the small pie plate. Do not soak the other two seeds.
3. Place enough paper towelling in the beakers to hold the seeds firmly between the paper and the side of the beakers.
4. Label the beakers 1, 2, and 3 with the marking pencil.
5. Place the unsoaked seeds between the towelling and the sides of beaker 1. Put the beaker in a spot where it will be in the light, and at room temperature for at least five days.

6. Place two of the soaked seeds between the towelling and the sides of beaker 2. Slowly add water to the beaker until the water is just below the seeds. Put beaker 2 beside beaker 1.

7. Place the remaining two soaked seeds between the towelling and the sides of beaker 3. Slowly add water to the beaker until it is just below the seeds. Put the box over beaker 3, and place it beside beaker 2.

8. Check the beakers each day for the next five days. Remember to cover beaker 3 each time. If necessary add enough water to beakers 2 and 3 to bring the water to the starting level. Observe and record any changes that may occur over the next five days. You may want to use a table like this one to record any changes you observe. If no changes are observed, write "no change."

Table 4

DAY	BEAKER 1	BEAKER 2	BEAKER 3
1			
2			
3			
4			
5			

Finding Out

1. (a) In your notebook list all the conditions for growth in beaker 1.
   (b) List all the conditions for growth in beaker 2.
   (c) List all the conditions for growth in beaker 3.
2. (a) What condition(s) did all three beakers have in common?
   (b) What conditions did beakers 1 and 2 have in common?
   (c) What conditions did beakers 2 and 3 have in common?
   (d) What conditions did beakers 1 and 3 have in common?
3. (a) Based on your observations, what conditions are necessary for seeds to begin growing?
   (b) How did your observations compare with your prediction?

PAPER TOWELLING
BEAN SEEDS
TOP OF WATER LEVEL
BOX OVER BEAKER 3

Seeds that Travel to Grow

Did you know that when you blow the seeds from a dandelion head, you are helping the plant? Seeds need space to grow. The dandelion, like many other kinds of plants, has ways to spread or **disperse** its seeds to other areas. Some seeds are carried by wind, some by water, and still others are dispersed by animals. For example, seeds of the snowberry bush are dispersed by many different kinds of birds living in the Canadian prairies. Here's what happens: a prairie bird, such as the sharp-tailed grouse, eats the white fruits of the snowberry bush; the juices in the bird's digestive tract break down the tough seed coats; the seeds go through the bird's digestive system, and the bird eliminates them. If the seeds land where the conditions are right for growth, a new snowberry bush will begin to grow.

The seeds of the snowberry will only begin to grow if:
(a) they have passed through the digestive tract of an animal like the sharp-tailed grouse;
(b) they are eliminated by the animal;
(c) they land in a place with good growing conditions.

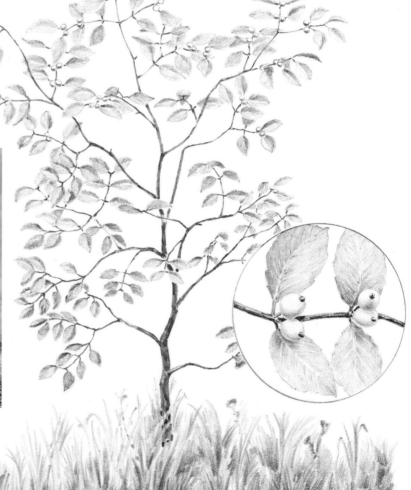

Look at the diagrams of these seeds of some common plants. By what means do you think each seed is dispersed?

beggars-tick

snowberry

cocklebur

dandelion

milkweed

maple

acorn

Did You Know?

Farmers find that weeds such as wild oats, sow-thistle, and wild mustard are the greatest plant pests of the Canadian prairies. Yet there was a time when these weeds were unknown on the prairies. The early settlers who came to farm brought bags of grain with them. Unknown to the farmers, the bags contained seeds of these weeds as well as grain seeds. At first the farmers ignored the weeds because there were very few of them. But the weeds increased quickly and today farmers have to spend thousands of dollars to control them.

The wild oat.

The Many Ways Organisms Reproduce

As you saw in Topic One, organisms have many different ways of reproducing. Some organisms need to mate, others merely reproduce on their own, as in the examples shown here.

These zebras rub against one another as part of the ritual that occurs before mating.

Two paramecia join together and exchange cellular material before splitting into four new paramecia.

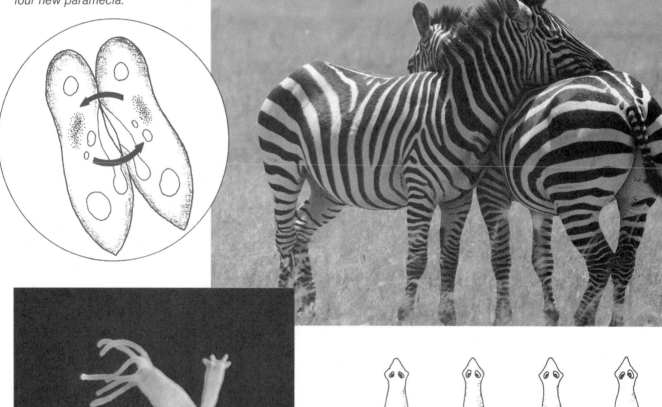

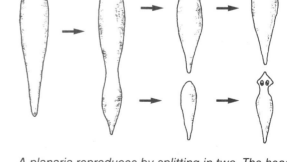

One way the hydra, a small pond animal, reproduces is by forming a "bud." The bud eventually breaks away and grows into an adult.

A planaria reproduces by splitting in two. The head grows a new tail, and the tail grows a new head.

Plants Are the Winners

Among living things, plants have the greatest variety of ways to produce more plants like themselves.

Probing

Roots can also produce new plants. Think of an example of a plant that reproduces in this way.

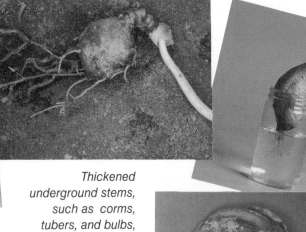

*After **pollination** by insects, wind, or water, plants produce seeds that grow into new plants. This syrphid fly is pollinating the flower.*

Thickened underground stems, such as corms, tubers, and bulbs, also produce new plants.

New plants grow from underground runners.

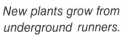

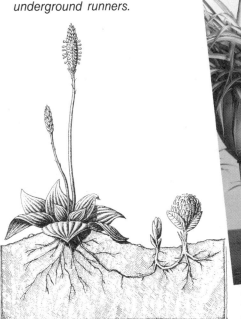

*New plants form on **shoots** or **runners**.*

*Leaf and stem **cuttings** will take root in water or moist planting material like sand.*

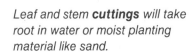

Growth and Reproduction **25**

Checkpoint

1. How well do you remember the meanings of words you have been reading? Copy the list of words in your notebook. Beside each, write the definition that describes each word. One word is not defined. Write a definition for that word.

(a) organism	a characteristic of life that allows organisms to change and repair injuries
(b) life cycle	
(c) characteristics	living things are made up of these tiny units
(d) growth	a living thing
(e) regeneration	the process of reproduction, the pattern of growth, and the life of the adult organism
(f) species	
(g) cells	identifying features
(h) photosynthesis	the ability of organisms to produce offspring that are similar to themselves
(i) reproduction	the ability to grow whole new body parts
	a group of living things in which all organisms are similar in size and shape

2. Narmutha's class was asked to classify several things as either living or non-living. Narmutha wrote down the following data about one particular thing.
   - It gets bigger.
   - It moves when touched.
   - It reproduces by splitting in two.

   Based on Narmutha's data, is the thing living or non-living? Explain which piece of information allowed you finally to classify the thing.

3. A puddle increases in size during a rainstorm. A baby robin grows bigger. Which statement is an example of a characteristic of life? Explain why you chose the statement you did.

4. In Topic Two, you read about various ways that organisms reproduce. Write the name of an organism that reproduces in each of the following ways.
   (a) A "bud" grows from a simple pond animal.
   (b) A plant reproduces with the help of a bird.
   (c) A plant reproduces from thickened stems.
   (d) A single-celled organism splits in two.
   (e) A plant reproduces from a runner.

5. Julia wanted to find out more about how plants grow. Here are the things she did. They are out of order. Arrange the actions in the order in which she did them.
   - She collected all the things she needed for the Activity.
   - She wrote a report about what she had learned from her observations.
   - She wrote down a prediction about conditions necessary for plant growth.
   - She recorded all her observations in a chart.
   - She carefully observed radishes growing under different conditions.

6. You have been given a new puppy. It has a large head and a short body.
   (a) You predict that the puppy will grow into a large dog. Explain what a prediction is and why you made this prediction.
   (b) You want to test your prediction by recording the puppy's growth over a period of one year. What measurements might you make each month?
   (c) You want to organize the measurements you make. Design a table to do this.

Built for Survival

All organisms obtain the things they need to survive, such as food and water, from their surroundings, or **environment**. The environment includes all the other organisms that live in the same surroundings. For example, the sharp-tailed grouse and the snowberry bush are both part of the environment of the Canadian prairies.

The environment also includes non-living things, such as air and water. The grouse obtains its water from small pools. The snowberry bush obtains its water from the ground.

This Topic is about the ways the body parts of an organism assist it to obtain the things it needs from the environment. Special body parts that help the organism to survive in its environment are called **structural adaptations**. For example, the giraffe's long neck allows it to eat leaves that grow high up in trees. The snowberry bush has roots to obtain water from the ground. The giraffe's long neck and the roots of the snowberry bush are both structural adaptations.

Birds have wings, trees have trunks, and turtles have shells. These are three examples of structural adaptations. How does each adaptation help the organism to live in its environment?

Structures for Getting Food

All living things need energy to grow, to reproduce, and to move. The source of this energy is food. Different organisms eat different kinds of food, and have different structures to obtain these various foods. Look, for example, at the insects shown on this page. All have six legs; all have the same way of breathing; all are similar in many other ways. But their mouthparts have become highly specialized to allow them to use very different food sources. Look at each of their structural adaptations.

Insects are not the only organisms that have highly specialized mouthparts to enable them to eat certain kinds of foods. In Activity 1-8 find out what structural adaptations some birds have to help them get their food.

Green plants make their own food by the process of photosynthesis. Special structural adaptations allow plants to obtain carbon dioxide from the air, and water from the ground. The plants change the carbon dioxide and water into food by using the sun's energy. Food that is not being used is stored for future use in the roots or stems.

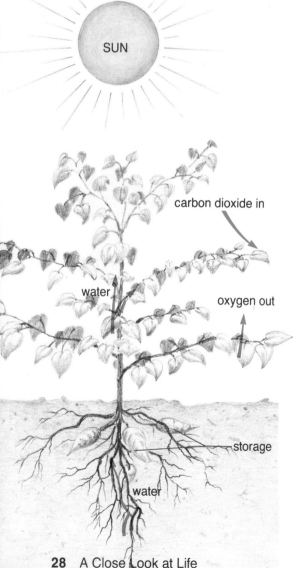

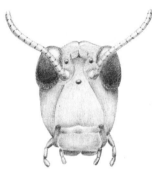

The grasshopper's mouthparts resemble teeth on a saw. These allow it to bite and chew the leaves and stems of plants.

The housefly's mouth is designed as a sucking tube with a wide end like a sponge, which it uses to draw up liquids from decaying food and waste.

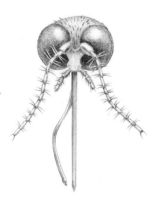

The mosquito's mouthparts are adapted for piercing and sucking. What is its food? How do its mouthparts assist it to obtain this food?

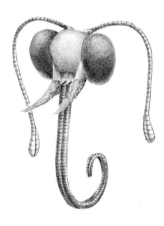

The butterfly's mouth is like a flexible straw, which it uncurls to sip nectar from flowers.

Prepared for the Job

In your notebook, match the bird to its food-getting habit by looking closely at the type of beak each bird has.

A. I fly low along open water, with my bill slightly open, and my lower bill in the water so I can skim small organisms right off the surface.

B. I use my strong jaw muscles and strange-looking bill to pry apart the scales on pine cones so I can get at the seeds inside to eat.

C. I breed in prairie sloughs. You'll see me there feeding as well. I wade in shallow water, swinging my bill back and forth in the water to gather tiny organisms swimming in it.

D. I have a straining device in my mouth. I take in water that contains tiny organisms, filter out the water, and leave the organisms inside my mouth. I tend to turn my head upside down to eat. By the way, I have a long neck and legs, and I'm pink.

E. You'd confuse me for a bee if you couldn't see clearly that I'm a bird. That's because I often eat the same food as bees—nectar from flowers. I also eat insects, although not very many people know that.

F. I fly along, ready to open my mouth wide as soon as I see an insect—an efficient way of catching insects in flight. Bristles around my beak help funnel the flying insects into my gaping beak.

G. I'm a diving duck. I have teeth-like structures on my beak to keep the fish I'm attacking from slipping away.

H. I'm a dabbling duck. I don't bother to dive. I just dip from time to time to eat floating vegetation. I often hang around ponds in city parks.

I. My beak serves the same purpose as a dog's canine teeth. I can rip and shred meat with it. Because I attack other organisms, I am called a "bird of prey."

J. I pound on trees to get to the beetle larvae, ants, and termites that live in the wood.

K. I am a close relative of the bird whose food is described just above this. Personally, I prefer sap from trees, and insects that come to drink the sap.

L. I feed along the shores of ponds and lakes and in sloughs in the prairies. I'm a large wading bird and have long legs so I can easily wade in water. To get food, I stand silently, suddenly stab at frogs and small fish with my sharp and pointed bill.

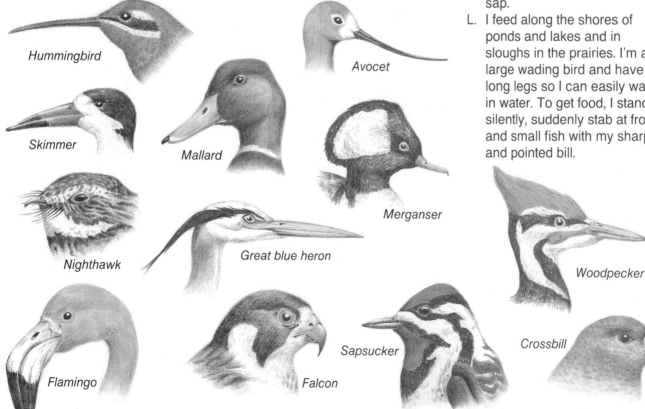

Hummingbird

Avocet

Skimmer

Mallard

Merganser

Nighthawk

Great blue heron

Woodpecker

Flamingo

Falcon

Sapsucker

Crossbill

Swallowing It Whole or in Part

Organisms that obtain food from other organisms must break the food down to get energy from it. The process of breaking food down into tiny usable particles is called **digestion**. Only after the food is digested can the organism pass the food particles to all parts of its body. The tiny particles of food supply the organism with building materials for growth and repair and with energy to perform other life functions.

Structures for the process of digestion can be very simple—or very complicated.

The single-celled amoeba simply surrounds its food with its body. A tiny bubble-shaped structure forms around the food as it is brought into the cell.

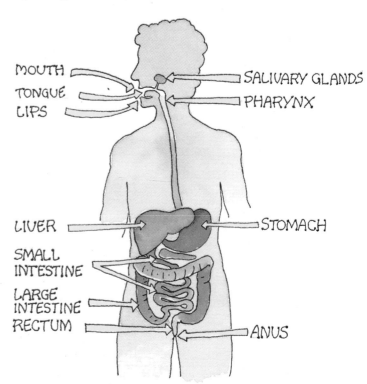

You have a very complicated way of digesting food. When you eat a sandwich, it passes through your digestive system, undergoing many changes before the energy in the sandwich can be used.

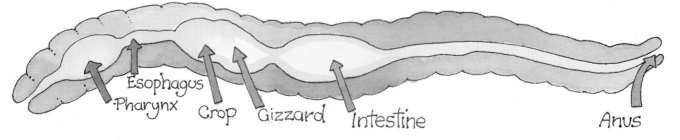

An earthworm's food is in the soil. As the earthworm moves forward, it eats the soil. The soil passes through the earthworm's tube-like digestive system. The food is removed from the soil before the soil is passed out of the earthworm's body.

Getting the Gases in and out

Like food, oxygen is a common need of most organisms, from tiny micro-organisms and plants to the largest animal or tallest tree. Oxygen is needed to change food into the energy needed for growth and movement. As well as taking in oxygen, all these organisms need to get rid of other gases, such as carbon dioxide. You have structures called lungs that enable you to breathe in air, which includes oxygen, and to breathe out carbon dioxide.

Activity 1–9

Checking Your Lung Capacity

Problem

How much air can you move in and out of your lungs in a single breath?

Materials

water
500-mL measuring cup
a plastic bucket
waterproof marking pen
a balloon or a plastic freezer bag

Procedure

1. Half fill a plastic bucket with water. Using a waterproof marking pen, mark a line on the bucket at the top of the water. Label this line 0 mL.
2. Pour 500 mL of water into the bucket. Use the marking pen to indicate the change in the water level. Label this line 500 mL. Add 500 mL more water. Use the same process to mark lines at 1000 mL, 1500 mL, 2000 mL, and so on until the bucket is nearly full.
3. Carefully pour out the extra water you added until the level of the water in the bucket is again at the level you marked as 0 mL.

4. Breathe in as deeply as you can. Breathe out as much gas as you can into a balloon or a freezer bag.
5. Quickly seal the open end of the balloon or bag with your finger or a piece of string.
6. Once the gas is securely trapped inside the balloon or bag, submerge it in the bucket of water. Try not to put your hand in the water.
7. Record the water level in the bucket.

> **CAUTION:** Dispose of your balloon or freezer bag after using it. Do not let anyone else use it.

Finding Out

1. (a) How far did the water level rise in the bucket after step 6?
   (b) What is your lung capacity?

Finding Out More

2. What effect would submerging your hand in the water have on the measurement of your lung capacity?
3. How could you change step 6 of the method you used so that you could use your hand to hold the balloon under water and *still* measure your lung capacity accurately?
4. Describe a situation in which it might be useful to have a large lung capacity.

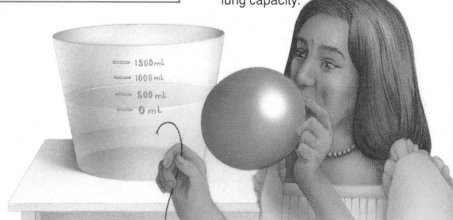

Organisms without Lungs

Many living things—frogs, snakes, alligators, birds, and some mammals—have lungs to take in oxygen. But lungs are not the only structure that living things have to get gases in and out. Take a look at these drawings.

Have you ever thought about how some organisms live in the same area but in totally different environments? Think about the tadpole and the frog. They both live in the same area, but one lives in water, and the other mostly lives on land. As the tadpole grows into a frog, its body parts change to enable it to survive on land.

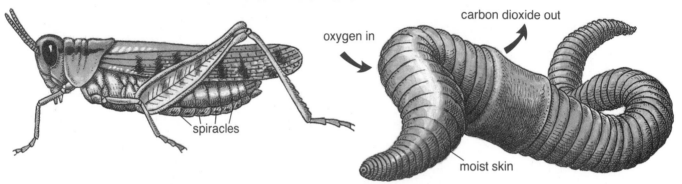

Grasshopper: Tiny openings called **spiracles** found on the grasshopper's sides allow gases to pass in and out of its body.

Earthworm: Oxygen passes in and carbon dioxide passes out through the earthworm's moist skin. If the skin becomes dry, the earthworm will be unable to get the oxygen it needs and will die.

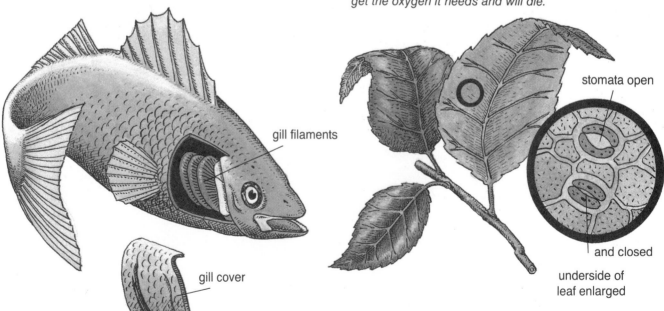

Fish: **Gills** on either side of the fish's head allow the fish to take oxygen from the water. Unwanted gases pass back into the water through the gills.

Plants: Tiny slits on the underside of a leaf, called **stomata**, open and close to allow oxygen and carbon dioxide to pass in and out.

1. Mammals are organisms with mammary (milk-producing) glands. In this way, all mammals are the same. But mammal species are very different in many other ways. Choose one of the mammals listed here. (Some mammals that are **extinct**—once living, but the species no longer exists—are included.) Find out about the environment of the mammal you have chosen and the kind of food it eats (or ate). Present its structural adaptations for obtaining its food in a poster or an illustrated report.

   sea otter
   sabre-tooth tiger (extinct)
   koala bear
   Przewalski's horse (a wild horse)
   beaver
   snowshoe hare
   caribou
   peccary (a wild pig found in Mexico and some parts of the southwestern United States)
   mountain sheep
   grizzly bear
   baboon
   kangaroo rat
   duck-billed platypus
   woolly mammoth (extinct)
   killer whale
   badger

2. This photograph shows a shallow pool. List three organisms that you think might live in or around the pool. For each, describe an adaptation that you think would assist the organism to survive in its environment.

A slough, or shallow pool.

3. People now travel long distances into space or under the ocean. Crews in spacecraft or submarines continue to eat and breathe normally. How has technology enabled people to live in these artificial conditions? Investigate one of the following: the mechanism on a submarine that supplies air to the crew; the mechanism on a space vehicle that supplies air to the crew; the mechanisms that provide food to the crew.

A Close Look at Movement

Look at either a clock or a watch that measures seconds. Time yourself. Try not to blink your eyes for thirty seconds. Can you do it? Why is it difficult?

We often think of movement as moving from place to place, or **locomotion**; but other kinds of body movements may be just as important. You blink your eyelids to keep your eyes moist. Blinking at just the right moment can protect your eyes against injury, just as moving out of the way of a falling rock can save your head from a nasty blow. This Topic investigates how living things move.

Animals with Legs and Skeletons

Let's start the study of movement with what's most familiar —the movement or locomotion of organisms that have legs. The human and grasshopper legs shown here look very different, but they both rely on similar structures for locomotion. Both have skeletons that provide support for their bodies, and legs to enable them to move.

The skeleton of a human supports the organism from inside its body. The skeleton needs to be strong and heavy because it has to support a large body mass. The skeleton of an insect is very different from that of a human. It forms a tube around the outside of the insect's body. The insect's skeleton is also strong: it not only supports the insect, but it also provides protection from animals that eat insects.

Organisms with skeletons have muscles anchored to their skeletons. This is true whether the skeletons are inside or outside their bodies. When a human or a grasshopper moves its legs, muscles in the legs contract (shorten). In this way the muscles pull or push the legs in the direction needed for locomotion. The animal moves forward as its legs push against the ground.

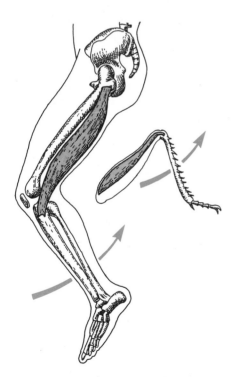

Human beings and grasshoppers both have legs—structural adaptations that enable them to move about on the ground.

Some insects fly faster than others. Can you use the information in the table to explain why some insects are adapted to move quickly while others are not?

INSECT	FLIGHT SPEED IN KILOMETRES PER HOUR (km/h)	WHAT THE INSECT EATS
Dragonfly	40	other insects
Honeybee	21	nectar from flowers
Housefly	19	many kinds of foods; mainly rotting material
Butterfly	8	nectar from flowers, or nothing at all in the adult stage

Did You Know?

Insects are the strongest jumpers of all animals. A flea can jump 50 cm! You're not impressed? Well, think about it. The flea is actually jumping 200 times its body length. A typical grade 7 student would have to jump more than the length of two football fields to equal this feat.

Locomotion in Animals without Legs

It may be easy to see how animals with legs move from place to place, but what about animals without legs? Think about snakes. Like human beings, snakes have skeletons inside their bodies. Attached to their skeletons they have muscles that can contract. But they don't have legs to push against the ground. How, then, do they move? The most familiar type of snake movement is shown in the diagram. As its muscles contract, first on one side and then on the other, the snake slithers along in a series of curves. It moves forward by pushing against sticks, rocks, grass, or even slight bumps on the surface of the ground.

In this diagram, the snake is using three rocks to help push itself forward.

The snake's scale-covered body helps the snake to grip the ground and stops the snake from slipping backwards as it winds its way from place to place.

No Legs and No Skeleton

Now let's consider an organism that seems to have a real problem—no legs and no skeleton. Yet the earthworm does perfectly well moving in and out of its burrow in the ground, and along the ground. How does it do it? Several things about an earthworm's structure are well adapted for its movement. (1) Its body is flexible because it is divided into segments. (2) Each of its body segments has two sets of muscles lying just under the skin. The top layer of muscles runs around the worm's body. Under this is another layer of muscles that runs along the worm's entire body from one end to the other. When the top muscle layer contracts, the worm becomes longer and thinner. When the inside muscle layer contracts, the worm becomes shorter and fatter. (3) Each segment of the worm has four pairs of bristles. The worm can push these out from the sides and bottom of its body to grip the ground as the worm moves forward.

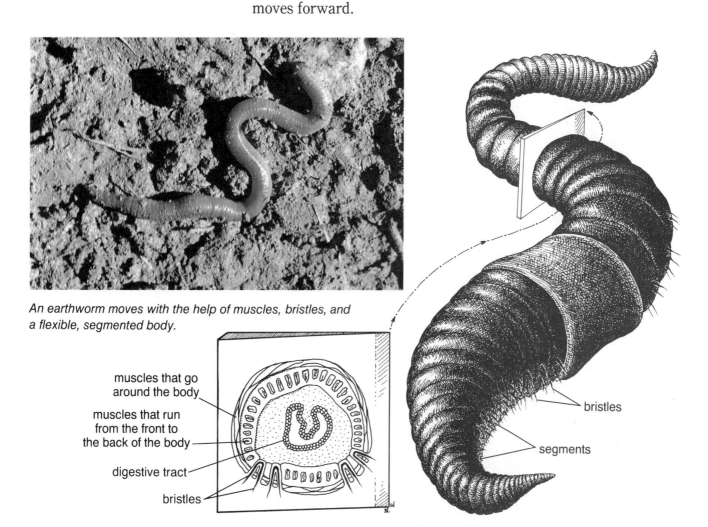

An earthworm moves with the help of muscles, bristles, and a flexible, segmented body.

muscles that go around the body

muscles that run from the front to the back of the body

digestive tract

bristles

bristles

segments

Care of Living Things

On a field trip you can see many living things, and watch some of them move about. In the classroom you can look at slides, movies, or videos of living things moving, growing, and reproducing. Sometimes you may be able to observe living things closely in the classroom. Before you do, keep this thought in mind:

Living things must be treated with respect.

If you have a pet, you know that you must give it food and water, keep it healthy, and treat it with kindness. Living things used in classroom activities must also be treated with kindness. Do not hurt them, and do not handle them in ways that would cause them stress. An earthworm may not react to pain in the same way that a cat or a horse might, but that doesn't mean that it should be treated any less carefully.

Make sure that plants and animals have the right conditions to survive—food, water, the right temperature. Whenever possible, return the organisms to their natural environment when you have finished studying them.

Treating organisms with respect means handling them with care.

Movement in Water

As you have seen, animals on land usually move by pushing some part or parts of their body against the solid ground. Fish move and must control their movements while being surrounded by a fluid—water—that flows around their bodies.

For a fish to move through water, it must push the water aside. If you are a swimmer, you will know that as you swim forwards, you push the water back with your arms. The fish uses its whole body to push the water back. It does this by curving its body, or wriggling back and forth.

Although the fish does not have arms, it does have other structural adaptations that help it move in water. The fish's fins and tail not only help it move forward, but they also control movement in other directions. Again, if you are a swimmer, you know that you are likely to sink in the water if you do not keep moving your arms and legs. And you must make small movements of your body in order to turn on your side or roll on your back. The fish has to make these kinds of movements just as a swimmer does.

If you have ever watched a small boat in rough water, you know that it can roll from side to side, or pitch from end to end. The fish has to use its body to control these kinds of movements too. If you have access to an aquarium, you could watch to see how a fish moves. Activity 1-10 gives you a procedure for doing so.

The Living Submarine

Problem

How do the various fins on a fish help it move?

Materials

one small goldfish taken from a large aquarium
a clean, large jar or beaker (about 1 L)
dip net

Procedure

1. Study the names of the fins shown in the illustration. (**Note**: It is not necessary for you to memorize these names. However, being able to refer to them will enable you to work through this Activity more easily.)
2. Fill a clean, well-rinsed jar or beaker with 600 mL of aquarium water.

3. Use the dip net to catch a goldfish in the large aquarium and carefully transfer the fish to the jar or beaker. Place your hand over the top of the container so that the fish cannot jump out.
4. Place the container on a flat surface and allow a few minutes for the fish to calm down.
5. While the fish is calming down, read through the Finding Out questions.
6. Study your fish for several minutes. In your notebook, list five to ten observations about the motion of the fish.

> **CAUTION:** Do not do anything that might harm the animal. When you have finished observing the fish, carefully return it to the aquarium.

Finding Out

1. (a) Which fins are used for moving forward? Explain how they help the fish to move forward.
   (b) Which fins are used for "steering"? Explain how they help the fish to steer.
2. How is the movement of the fins similar to, or different from, the movement of a swimmer's arms?
3. How does the movement of the fins compare with the movement of oars in a rowboat?
4. Did the fish go backward as you watched it? What movements did it make to do this?

Finding Out More

5. How is the flight of birds similar to the swimming of fish?
6. What flows around the body of a bird as it flies?

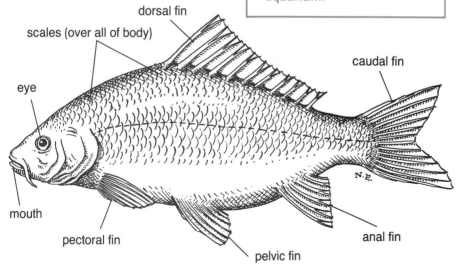

Fins of a typical fish.

A submarine is a ship that can travel underwater in much the same way as a fish can. A submarine is able to submerge, or sink, by allowing water to flood into ballast tanks inside the hull. The submarine rises to the surface again by pumping water out of the ballast tanks. Controls within the submarine allow the navigator to prevent the submarine from pitching or rolling from side to side, or even rolling over. The rudder is used to keep the submarine moving straight through the water, or to turn it in the desired direction.

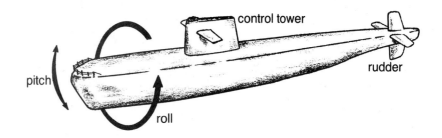

Movement that's More Difficult to See

You control the type of movement we call locomotion by deciding to walk, run, and so on. You also control other kinds of body movements such as laughing, or swallowing when you are eating. You control some other kinds of movement in your body only when you are thinking about them—remember how difficult it was to control blinking your eyes. There are still other kinds of body movements that you do not seem to control at all. These body movements are often not easy to see.

When people are sick, physicians and nurses keep a check on the body's **vital signs**. Vital signs include breathing rate, body temperature, pulse rate, and blood pressure. Some of these signs involve body movements that can be measured.

Each time the heart muscle contracts to pump blood out of the heart to the rest of the body, it produces a rhythmic pulse. The pulse is fairly easy to detect in most people.

Your Pulse Rate

Problem

What is your pulse rate?

Materials

stopwatch or clock or watch that measures seconds

Procedure

1. Find the pulse in your wrist: place the arched middle fingers of one hand between the bone nearest your thumb and the ridge in the centre of your wrist on your other hand. Gently press down until you feel your pulse. Do this several times until you know where to find your pulse beat. If you have trouble finding the pulse in your wrist, use the middle fingers of one hand to find the pulse in your neck instead.

Table 5

TRIAL	NUMBER OF PULSES (in 30 s)		NUMBER OF PULSES (in 1 min)		
1		× 2 =			
2		× 2 =			AVERAGE NUMBER OF PULSES (in 1 min)
3		× 2 =			
	TOTAL OF TRIALS =			÷ 3 =	

2. Make a table like the one above to record your observations.
3. Find your pulse and count the number of pulses you feel in exactly 30 s (seconds). Record your results in your table as "Trial 1."
4. Repeat the count two more times. Record your results.

Finding Out

1. Multiply each count by 2 to find the number of pulses in 1 min.
2. Find your *average* pulse rate by adding the numbers in the column "Number of pulses in 1 min," and then dividing the total by 3.
3. Record the average pulse rate of each student in a table on the chalkboard. As a class, work out how to find the average pulse rate for your class.
4. You will probably find that there is a fairly large range in the data you have collected.
   (a) What is the range of pulse rates in your class?
   (b) Suggest some reasons why pulse rates are not the same for everybody in the class.

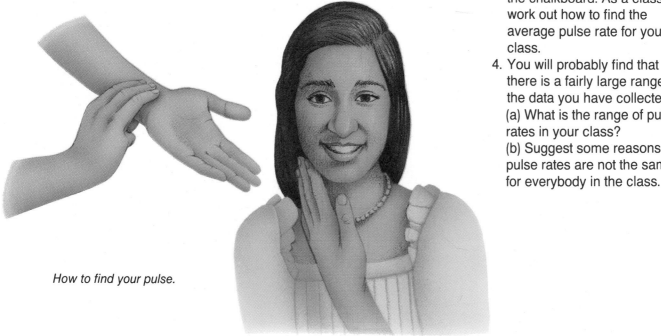

How to find your pulse.

The Effect of Exercise on Pulse Rate

Problem

How does exercise affect pulse rate?

Define the problem.

(a) Does exercise have *no effect* on your pulse rate?

(b) Does exercise cause your pulse rate *to increase*?

(c) Does exercise cause your pulse rate *to slow down*?

> **CAUTION:** If for any medical reason you should not run on the spot, act only as a recorder in this Activity.

Procedure

1. In your notebook, write a prediction about the effect of exercise on your pulse rate. That is, decide which of (a), (b), or (c) above you think is most likely.

2. Now test your prediction by an experiment. You will need a **control**, or standard pulse rate against which to measure your pulse rate after exercise. That control is your average pulse rate, which you measured in Activity 1-11. (Read about controlled experiments in *Skillbuilder Two*.)

3. In your experiment you will use *time* as your **manipulated variable** (the variable that you use to measure your pulse). You will measure and record the **responding variable**, that is, the change in your pulse rate after you exercise.

4. Before you begin, draw in your notebook a table similar to the one you used for Activity 1-11. On the table record the data you obtain from this experiment.

5. In order to compare your data from this Activity with the control data from Activity 1-11, your experimental procedures should be similar.

(a) Run on the spot for 15 s.

(b) Count the number of pulses you feel in exactly 30 s.

(c) Record your results in your table.

(d) Rest for at least 2 min to allow your pulse to return to its normal rate.

(e) Repeat steps (a), (b), (c), and (d) twice more.

Finding Out

You have obtained your data and organized them in a table. Now *analyse* your data.

1. Find your average pulse rate after exercise.

2. Now you are ready to *interpret* your data. How did exercise affect your pulse rate?

3. Did your experimental data support your prediction? Explain.

4. Using the data you obtained in this Activity and in Activity 1-11, draw a bar graph of your average pulse rate before and after exercise.

5. What conclusions can you make about the effect of exercise on your own pulse rate?

Finding Out More

You now know the effect of exercise on your own pulse rate. But you cannot conclude that exercise will have the same effect on the pulse rate of other people without more *evidence*, that is, data from many more similar experiments.

6. Record the average pulse rates of each student before and after exercise in a table on the chalkboard.

7. What conclusion can you now draw about the effect of exercise on pulse rates of people in general?

8. You have solved a problem by following the same thinking and experimental process that scientists often use. You will be using this process again to design you own experiments. Write a summary of this process.

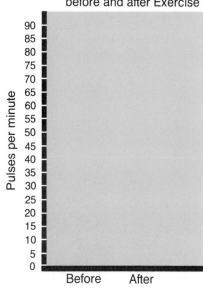

Average Pulse Rate before and after Exercise

Did You Know?

The speargrass seed has a pointed tip with hairs that anchor it to the ground. It drills itself into the ground by twisting.

Movement in Plants

Strawberry plants and many grasses send out shoots that then become new plants. These new plants also send out shoots. In this way the plant spreads across the ground. The original plant does not move, but some of its parts do—through growth.

Most movements in plants are less obvious than movements in animals. For example, plant movements are usually "movements in place," such as the bending of a plant toward the sun, as in the plant on page 5. Plants have no nerves or muscles as do the snake or earthworm. Just how, then, do plants make their movements?

Scientists have been curious about how plant movements occur, and have done many experiments to find out about such movements. How some plants move is still unclear. However, scientists have determined how the mimosa, or "sensitivity plant," suddenly droops its leaves when touched.

They found that when the leaves of a mimosa plant are touched, the plant immediately loses water from the base of each tiny leaf and from the point where the stem of the leaf is attached to the main stem. This causes the leaf to droop, because the water helps to keep the leaf rigid.

The walking fern is a good name for this fern. It sends out shoots that produce new ferns "walking" across an area from year to year.

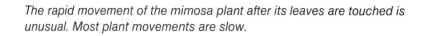

The rapid movement of the mimosa plant after its leaves are touched is unusual. Most plant movements are slow.

Checkpoint

1. In your notebook, unscramble the following letters to form words you've read about in Topics Three and Four. Then write a short description of each term.
   (a) NIRVENTOMEN
   (b) LARUCTRUST PADANTOTISA
   (c) LAVIT GINSS
   (d) ISTONGIDE
   (e) MOOTICNOOL
   (f) LUPES
   (g) VEMEMONT

2. Describe how each of these organisms moves from place to place: a snake, a grasshopper, and an earthworm.

3. In Topic Three, you learned that an organism may have one or more adaptations that help it to survive in its environment. For instance, a hummingbird has a long, tube-like beak for collecting nectar from flowers. For each of the following, name an organism and the adaptation that helps it in the situation described:
   (a) an animal that feeds on leaves high on trees
   (b) an insect that feeds on nectar from flowers
   (c) a bird that feeds on insects as it flies
   (d) a bird that feeds on larvae or termites in the wood in tree trunks and branches
   (e) an animal that breathes in water

4. Use examples to explain the following statement: "As some organisms grow, they go through many structural changes that help them to survive in their environment."

5. Which living things do you think influenced the inventors of these devices?
   (a) rubber flippers for skin-divers
   (b) kite
   (c) helicopter

6. Give an example of plant movement that shows
   (a) a plant changing shape;
   (b) a plant changing position;
   (c) a plant spreading across the ground.

7. Describe an everyday situation or task that involves changing the shape, position, and location of your body. Explain when each type of movement would occur as you do the task.

8. You have found that exercise has an effect on pulse rate.
   (a) Predict whether exercise would have a greater or lesser effect on the pulse rate of an athlete in training than it does on an average person.
   (b) Design an experiment to test your prediction.
   • State the problem or question you want to investigate.
   • Record your prediction.
   • Write down what you would use as the control in your experiment.
   • Write down what the manipulated and responding variable in your experiment would be.
   • Design a method of recording the data from the experiment.
   • Write down the steps in the procedure you would use.
   • Explain what you would do to analyse your data.
   • Explain what you would do to interpret your data.

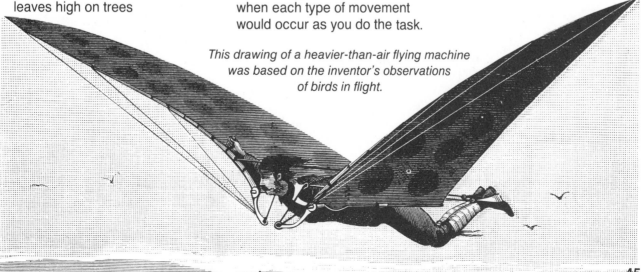

This drawing of a heavier-than-air flying machine was based on the inventor's observations of birds in flight.

Sensing and Responding to the Environment

The boy, the cat, and the birds in the drawings are **responding** (reacting) to an important change, or **stimulus**, in their surroundings. A single stimulus may cause a series of responses in an organism. You are riding your bicycle and see a red light ahead of you. You respond to the red light by braking, looking over your shoulder to check on other traffic, steering your bike to the side of the road, moving your foot from the pedal to the pavement to steady yourself, and waiting for the light to turn green. Each of these responses to the stimulus of the red light helps you to survive.

Each organism has specially adapted body parts through which the organism receives information about its environment. You receive **stimuli** (the plural of stimulus) through your eyes, ears, nose, skin, and tongue. Each of these structures, or **sense organs**, is associated with a different **sense**. Each receives only one kind of stimulus; for instance, eyes detect only light.

The speed of an organism's reaction time—the length of time it takes the organism to respond to a stimulus—could save its life. How quickly can you react to a stimulus? Do Activity 1-13 to find out.

Reaction Time

Problem

What is your reaction time?

Materials

metre stick or a piece of stiff cardboard marked in centimetres
pencil
paper
a partner

Procedure

1. Read through the steps. Then estimate how far you think the metre stick will drop before you catch it. Record your estimate.
2. Rest your arm on the desk. Have your partner hold the metre stick with the 0 cm mark just above your open thumb and forefinger as shown.
3. Keep your eye on the metre stick. Your partner will drop the metre stick without warning.
4. Catch the metre stick by closing your thumb and forefinger around it. Do not remove your arm from the desk as you catch it.
5. After you catch the metre stick, read the number (in centimetres) just above your closed fingers. Record this number on a copy of Table 6. The distance the metre stick dropped before you caught it indicates the speed of your reaction. The longer the drop, the slower your reaction time.
6. Before you try again, predict whether or not you think practice will improve your reaction time.

Keep your eye on the stick, and your arm resting on the desk.

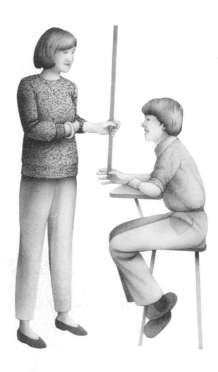

Table 6

TRIAL NUMBER	DISTANCE STICK DROPS IN CENTIMETRES (cm)	
1		
2		
3		
4		
5		AVERAGE DISTANCE STICK DROPS IN CENTIMETRES (cm)
SUM OF DROPS	÷ 5	=

7. Repeat steps 1 through 5 four more times, recording the distance the metre stick drops each time.
8. Add the results of the five trials and divide by 5 to find the average distance the stick drops.
9. Change positions with your partner and repeat the Activity.

Finding Out

1. How close was your estimate to your first trial?
2. Did practice improve your reaction time?
3. How accurate were your measurements of the distance the metre stick dropped? What might have affected your measurements?
4. Why do you think some people have faster reaction times than others?

Finding Out More

5. Is being able to catch a falling metre stick important to the survival of a human being? Try to think of one or more instances where speed of reaction time would be important for survival.

Extension on next page

Receiving the Message

Extension

6. The reaction times in the previous Activity depended on the sense of sight to detect the falling metre stick.

• Suppose you were blindfolded, and had to rely on sound instead of sight. Do you suppose you would react to the falling stick faster or more slowly?

• What might happen to your reaction time if your partner touched your head lightly at the same time as dropping the metre stick?

(a) Design and carry out an experiment that will help you answer one of these questions. To help you design your experimental procedure, read through the steps in Activity 1-13. In your notebook, write down what changes you need to make to measure the new variables in the experiment.

(b) What data will you use as the control in your experiment?

(c) Carry out your experiment.

(d) How did your observations compare with the observations you made in Activity 1-13?

(e) What interpretations can you make from your data?

When one of your sense organs receives a stimulus, it does so because of a body part called a **receptor**. Receptors are located throughout your body. Each receptor is specially adapted to receive only one kind of stimulus, such as light, taste, or smell. The most numerous ones are found in the eyes, nose, ears, tongue, and skin.

From the receptor a signal is passed along through a series of nerve cells to your brain. The brain receives the information, analyses it, and sends back instructions to the body to respond to the stimulus in a particular way.

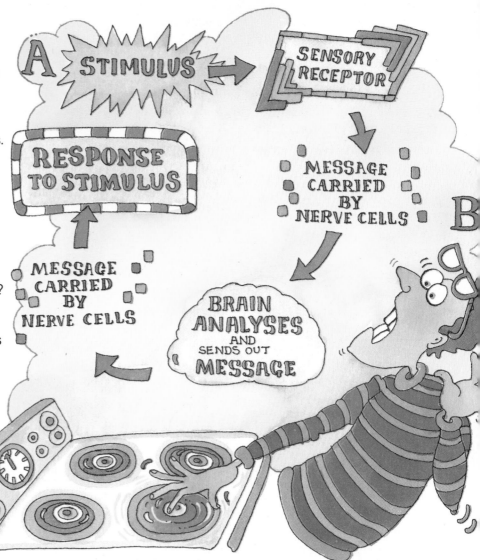

The hot element on the stove is the stimulus. Starting at A in the diagram, follow the steps that will take place before Dan responds to the warmth he feels as he brings his hand close to the hot element.

(a) Think of something you do every day, such as making a sandwich or tying your shoelaces. Which senses do you use to complete this task?
(b) Now imagine how you would do the same task if you had lost one of your senses, such as sight. On which sense or senses might you rely instead?

Did You Know?

Specially designed equipment helps impaired people to do things that, only a few years ago, would have been much more difficult for them.

A hearing-impaired person can receive telephone messages on the screen and send messages on the keyboard of this visual telephone.

This blind physiotherapist takes a patient's blood pressure using a device that gives an audible, rather than a visual reading.

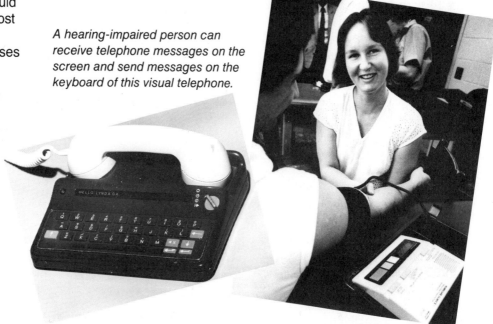

Did You Know?

Different parts of the brain control different body functions. The human brain can receive up to 100 million signals each second!

The Senses Our Lives Depend On

People can lose one of the senses that receive stimuli from their environment and continue to live successfully. There are some senses, however, that we cannot lose and continue to survive.

You breathe in and out continuously. You don't even have to think about breathing, your body just does it automatically. If you stopped breathing, you would not survive. What keeps you breathing? You've probably guessed it. There are sense receptors inside your body that detect *internal* stimuli. For example, you have receptors in your body that sense changes in the amount of carbon dioxide in your system. When these receptors sense that there is too much carbon dioxide, they send a message that causes certain muscles in your chest to contract. These muscles force carbon dioxide out of your lungs, that is, they cause you to breathe out.

Check your breathing rate. Your breathing rate is the number of times you breathe in and out (inhale and exhale) each minute.

Breathing Rate

Problem

What is your breathing rate?

Materials

pencil
paper
stopwatch or watch with second
hand

Procedure

1. Sit quietly, breathe normally, and count the number of times you inhale and exhale in 30 s.
2. Count "one" for each time you complete a full breath in and out. Stop counting after exactly 30 s. (Note: If you breathe in but not out when 30 s is up, count half a breath or 0.5.)
3. Record your data in a table like the one below.

4. Repeat steps 1 and 3 three times and record your results.
5. Total the time under the column "Breathing rate for 1 min." Find your average breathing rate.
6. Record the data about your average breathing rate along with that of your classmates in a table on the chalkboard.

Finding Out

1. Was there much variation in breathing rate among your classmates? If so, what might be some reasons for the variation?
2. You may have affected your breathing rate to some extent simply because you were thinking about it. Design a way to test your breathing rate that would avoid this problem. Discuss your ideas with your classmates.

Extension

3. In Activity 1-12, you discovered that your physical activity affected your pulse rate. Pulse rate is an indication of how quickly the heart is pumping blood to the body.
(a) Does physical activity also affect your breathing rate? That is, does physical activity affect the rate at which your lungs take air in and out? Are pulse rate and breathing rate connected?
(b) To find out, carry out the procedure as in Activity 1-12, but this time measure and record your breathing rate after exercise instead of your pulse rate after exercise.
(c) Draw a bar graph of your average breathing rate before and after exercise.
(d) Compare this graph with the graph of average pulse rates you drew in Activity 1-12. Is there a similar trend or pattern in each graph?
(e) Compare your results and your interpretations of them with your classmates' results.

Table 7

TRIAL NUMBER	BREATHING RATE (in 30 s)			BREATHING RATE (in 1 min)	
1		× 2	=		
2		× 2	=		AVERAGE BREATHING RATE (in 1 min)
3		× 2	=		
	TOTAL		=	÷ 3 =	

A World Filled with Stimuli

You need light to see what you are doing, whether it is light from the sun or light from devices that people have invented. But you usually sleep in the dark. Different organisms react in different ways to stimuli such as light. A plant needs the sun's energy to produce food. The plant grows toward light—it responds *positively* to light. An animal that lives in a burrow underground comes out only when it's dark. Such an animal tends to show a *negative* response to light.

How do other organisms respond to different stimuli? Do Activity 1-15 to observe the responses of one kind of organism to a number of stimuli.

YOU CAN LIVE FOR **WEEKS** WITHOUT FOOD

DAYS WITHOUT WATER

BUT ONLY **MINUTES** WITHOUT OXYGEN

Tulips respond positively to the sunshine, opening on a sunny day, closing at night, and staying closed during an overcast day.

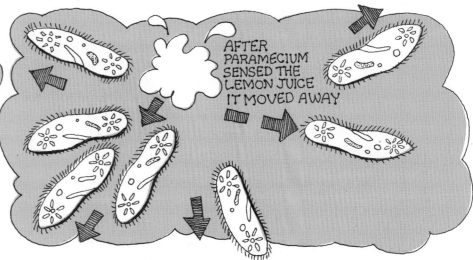

AFTER PARAMECIUM SENSED THE LEMON JUICE IT MOVED AWAY

Acid is harmful to these tiny, one-celled pond creatures, and acts as a stimulus to them. You can see how they move away from an area where a weak acid has been added. This negative response to a drop of acid is important to their survival.

Meandering Mealworms

Problem

How do mealworms respond to various stimuli?

Materials

2 or 3 mealworms
spoon
index card
shallow tray or shoebox lid
paper towel
piece of paper
2 cotton balls
water
sheet of white paper
sheet of black paper
stop watch or watch with second hand

CAUTION: Treat the mealworms gently—they are quite fragile. When you wish to remove a mealworm from the jar, use a spoon. Gently roll the insect from the spoon into your shallow tray. Then lift and carry the tray to your work area.

Procedure

1. Place a mealworm on a piece of fairly rough paper towelling in your tray or upturned shoebox lid. Time how long it takes the worm to move 2 cm.
2. Now predict whether it will travel faster on a smooth surface. Test your prediction by placing the same mealworm on a smooth surface such as an index card.

3. Predict whether the mealworm will travel up or down a slanted surface, and whether the amount of the slant will affect its movements. Test your predictions by carefully propping up one end of the index card with a pencil or your finger.
4. Gently touch the mealworm with your finger. Record its response.
5. Change mealworms. Carefully return the first mealworm to its jar.
6. Predict whether a mealworm likes moisture or not. Test your prediction by placing the mealworm between two cotton balls about 2 cm apart. One of the cotton balls should be wet, the other dry.

Finding Out

1. Did any of the mealworm's responses seem to be ones that you would *not* expect from an animal that lives in decaying grain or flour? If so, why do you think it might have responded the way it did?
2. Explain how each of the mealworm's responses might help it survive in its environment.

Finding Out More

3. Predict a mealworm's response to darkness and light. Using sheets of white and black paper, design an Activity to test your prediction.

4. Review all your observations of a mealworm's responses to various stimuli. Use these as a guide to write a short paragraph or poem from the mealworm's point of view.

Extension

5. You may wish to observe the responses of another kind of organism. Garden snails, beetles, and earthworms are also interesting animals to observe. You could see how one of these organisms responds to light, touch, or other stimuli that will not harm the animal. When you are planning these activities, you should first find out more about the organism's environment. This will help you decide what tests to make regarding the organism's response to certain stimuli within its environment.

Caring for Mealworms

A mealworm is the larval stage in the life cycle of a grain beetle. This insect is usually found in rotting grain or flour supplies. Mealworms may be bought cheaply at most pet stores.

Mealworms are easily kept for fairly long periods of time in a glass jar that has air holes punched in the lid. Bran and other cereal grains are good foods for mealworms. Keep the cereals slightly moist by putting a slice of apple or potato into the jar. Change the mealworms' food when the bran looks powdery.

The mealworm is quite harmless, and if you care for it for six to nine months, you can observe its entire life cycle—egg, larva, pupa, and adult. Only the egg is difficult to see because it is so tiny.

Behaviour for Survival

Living things respond to their environment in ways that help them survive. These ways are called **behavioural adaptations**. Recall how you react when you are riding your bike and see a red light. Your series of responses helps you to survive the dangers of the busy street. Your behaviour is adapted to the many conditions in your environment.

It is fairly easy to see how your behaviour helps you cope with riding your bike in traffic. Behavioural adaptations in other organisms are not always so simple to understand. **Biologists**, or scientists who study living things, have done many experiments to try to understand behavioural adaptations in living things.

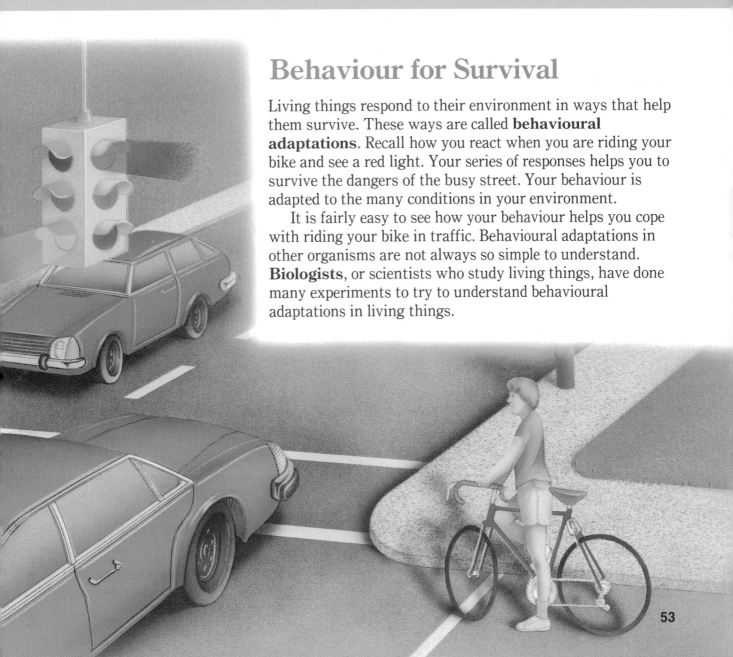

How Do Organisms Find Their Way About?

The famous biologist, Niko Tinbergen asked, "When organisms leave their nests or burrows, how do they find their way back again?" He designed an experiment to find out how digger wasps find their way back to their nesting holes. He thought that they probably memorize landmarks just as most humans would. He then carried out the following experiments to test his idea.

Dr. Tinbergen concluded that a digger wasp finds her way back to her nesting hole by memorizing the landmarks around the nesting hole. In this case, the landmark she used was the ring of pine cones.

Wasp Ways

1. Study the illustrations carefully.
2. In (a), why do you think the digger wasp did not fly away immediately when she came out of her nest?
3. What test was the researcher making in (b)?
4. Why was the test in (b) repeated several times?
5. What was being tested in (c)?
6. Do you think these experiments show that the digger wasp memorizes the shape of the landmark, or the kind of material the landmark is made of?

(a)

(a) A ring of twenty pine cones was put around a nesting hole while the female digger wasp was inside. When she left the nest, she buzzed around the area for six seconds, and then flew away.

(b)

(c)

(b) While the wasp was away, the researcher moved the pine cones, placing them in a ring about a half metre away. When the wasp returned after about ninety minutes, she went directly to the ring of pine cones, looking for her nesting hole. When these tests were repeated, the digger wasp flew to the middle of the pine cones every time.

(c) In another set of experiments, pine cones were again placed in a ring around the nesting hole while the digger wasp was inside. After she left the nesting hole, pine cones were placed in a triangle around the nesting hole entrance. A ring of 20 pebbles was placed nearby. On her return, the wasp went each time directly to the middle of the ring of pebbles, looking for her nesting hole.

How Do Chicks "Ask" for Food?

Many newly hatched birds are fed in the nest by their parents. The parent bird carries the food to the nest. When a parent bird arrives at the nest, the baby birds tap the parent's beak. This action causes the parent to plunge the food into the babies' gaping beaks.

Scientists wondered what stimulus might cause herring gull chicks to tap the parent's beak.

1. The chicks might be responding to a spot on the parent's beak.
2. The chicks might be responding to a spot of a particular colour on the parent's beak.

To test these ideas, scientists made five models of a herring gull's head and beak. Each model was shown to baby herring gulls.

Activity 1-17

How to Get Fed if You're a Herring Gull Chick

1. Study the reactions of the chicks to the five models.
2. Do you think the gull chicks can distinguish one colour from another? Explain your answer.
3. Why was model E included in the test?
4. What do you conclude is the most important stimulus to make young gulls tap the parent's beak?

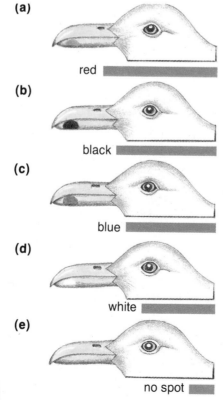

(a) red

(b) black

(c) blue

(d) white

(e) no spot

Five models of a herring gull's head were shown to baby herring gulls. The bars beneath the models show the number of times the baby gulls tapped each of the model heads. The longer the bar, the more times that model was tapped.

How Do Male Sticklebacks Recognize Each Other?

The male stickleback fish normally has a silver belly. During the mating season, the male's belly flushes red. When the male stickleback chooses a mate, he will fiercely attack any other male stickleback that comes near her. Scientists wondered what stimulus would cause this behaviour in male sticklebacks. Could it be just the sight of another male? Or was it something to do with the red colour of a male's belly?

To find out, four models were shown to male sticklebacks during the mating season. One looked exactly like a stickleback in shape and size, but had a silver belly. Three other models had red undersides, but none of these models was shaped like a stickleback. The male attacked models (b), (c), and (d), but not model (a).

(a)

(b)   **(c)**

(d)

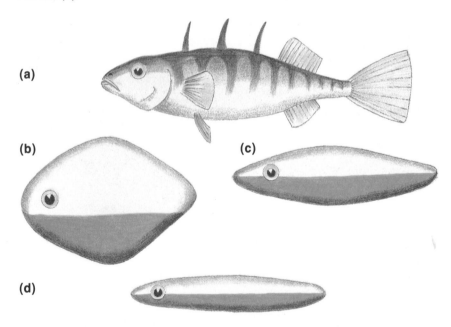

Activity 1-18

Stickleback Stimuli

1. Which stimulus do you think causes one male stickleback fish to attack another?
2. Why might a male stickleback try to keep other males away from his mate?

Instinct and Learned Behaviour

Some behaviour is **instinctive**. That is, an organism is born with the ability, or instinct, to respond automatically to a particular stimulus in a certain way. The organism does not change its response with experience and it does not have to be taught how to respond by another organism. The responses of the baby gulls and the male sticklebacks described on pages 55 and 56 are instinctive.

Other behaviour, particularly in vertebrates (animals with backbones), is **learned**. The digger wasp showed a very simple form of learned behaviour—an ability to find her way home by memorizing landmarks. This learning came from her own experience. However, most behaviour that is learned comes from imitating a model, such as a parent. If the organism does not have a model to learn from, it cannot respond to certain stimuli as do other members of its species. For example, you have probably heard the songs of several kinds of songbirds. **Biologists** (scientists who study living things) have found that these birds communicate by their songs. Experiments have shown that a songbird like a robin or a meadowlark must learn the song its species sings. If a baby bird is isolated from others of its kind and it does not hear its species' song early in its life, it will never sing in quite the right way. It will not be able to communicate effectively with others of its kind. That means it probably will not attract a mate and therefore will not produce young.

Songbirds, like this western meadowlark, must learn the song they sing.

Seasonal Behaviour

Organisms are adapted to survive in their environment, but what happens when the conditions in that environment change? The behaviour of each kind of organism must be adapted to these changing conditions. For example, your behaviour is adapted to seasonal changes from winter to summer—you wear different clothes. Other organisms have different responses.

Some animals stay in Canada all year long. Others live in one place during one season of the year, and move, or **migrate**, to another place for another season of the year. Some animals that live in Canada from spring to fall migrate before the winter to places south of Canada, where it is warmer.

It is only natural to think that the reason animals migrate is to avoid the cold of the Canadian winter. But if some animals avoid the cold in this way, why don't all Canadian animals migrate? Biologists have studied the behaviour of different kinds of animals to find the answer to this question.

Table 8

SMALL BIRD	DOES MIGRATE	DOES NOT MIGRATE	FOODS IT EATS
Black-capped chickadee		*	seeds, and insect larvae in bark of trees
Yellow-rumped warbler	*		flying insects
Red-breasted nuthatch		*	seeds, and insect larvae in bark of trees
Least flycatcher	*		flying insects
American robin	*		berries, moving insects, and worms

The chickadee and nuthatch are so small, you would think they would freeze to death in winter. But these birds do not migrate. They fluff up their feathers to keep warm, and remain active all winter in Canada. They eat seeds and insect larvae they find in trees. Now look at the kinds of foods the other birds eat. Do you see a pattern? The birds that eat moving insects and berries cannot find these foods in Canada in winter. They migrate to places where they can find such foods. They return to Canada in spring when these foods are available again. Therefore, it appears that they migrate south from Canada before winter because they would not be able to find the right kind of food during the winter.

The feeding behaviour of those small birds answered the question: "Why some small birds migrate and others do not." But biologists think other kinds of organisms migrate for other reasons. These photographs and diagrams show the migration patterns of several other well-known migrators.

Studying the behaviour of organisms can tell us a lot about why they act in certain ways at certain times. It can also help us understand our own behaviour better.

When conditions near the surface of the ground are either too dry or too cold for them, earthworms move down their burrows deep into the soil. When would you expect to find earthworms easily? Why do you think the graph starts with 0 cm at the top rather than at the bottom?

Earthworm migrations

*Blue whales spend the summer months in the Arctic building up their **blubber**, or insulating fat layers, by feeding on microscopic organisms (plankton) found in those waters. They do this before they are driven south by the spread of pack-ice as winter approaches. During the winter months they eat very little, living for the most part on the energy stored in the blubber. The young whale calves are born in the warmer southern waters, and are fed by their mothers until they are able to find their own food.*

Science and Technology in Society

Sharing Their Air Space

The sight of a bird soaring through the air makes many people think of freedom. Yet the sky is no longer as safe a place for birds as it used to be. Birds now have to share the air with people and their structures—buildings that are many times higher than any tree, and massive aircraft that thunder through the air.

Birds and Air Traffic Control

Birds in flight were the original inspiration for the airplane. However, birds and airplanes cannot share airports—at least not safely. Collisions between birds and planes are called strikes. Strikes nearly always kill the birds, but they can be fatal for the passengers and crew of the plane as well. The force of a collision may shatter a plane's windshield. Birds sucked into a jet engine cause a sudden loss of power: when this happens, the plane may crash. In one year, bird strikes cause an estimated 3 million dollars in damage to planes in Canada.

Air traffic controllers talk to the pilots of planes to keep incoming and outgoing planes from crashing into each other, but they can't talk to the birds. Some airports are now using birds of prey such as falcons to keep small birds from crashing into planes. All a falcon's trainer—the falconer—has to do is release the falcon into the air several times a day, being careful not to fly the bird too close to moving aircraft. When small birds detect the sight or shadow of a circling falcon, they either stay on the ground or flee.

Falcons were tried out at one of Canada's busiest airports. The results were impressive: the airport had far fewer strikes than other, less busy airports. But because some kinds of falcons are rare, we need to be sure they are used for purposes such as these only if there is no effective alternative.

Skyraker is one of a team of birds of prey used to keep gulls and other birds away from the runways of this busy Canadian airport.

A Deadly Weekend

When a new pair of tall smokestacks were built, people in the city were concerned that small aircraft might collide with the stacks. The solution seemed straightforward: aim powerful floodlights to light the stacks all night long. Then, one foggy Labour Day weekend, ten thousand small songbirds crashed into the stacks and died. Among the casualties were some species of birds whose numbers were already dangerously low.

What had happened? The smokestacks were on a migration route used by

During the fall migration, the lives of many birds have been saved by turning off the lights on the north side of very tall buildings each night. Why do you think this simple practice works?

birds. The night-flying birds had been attracted by the strong light reflecting off the smokestacks.

A way had to be found to warn aircraft of the stacks yet not attract the birds. The solution was to replace the floodlights with strobelights. Strobelights flash on and off quickly and so do not lead birds to the stacks.

Homes Can Be Hazards for Birds

Have you ever had a bird crash into the window glass at home? The glass acts as a mirror, reflecting the outdoors and causing the birds to fly straight at the window. A collision like this can kill a bird or knock it unconscious so that it becomes an easy prey for a cat or other predator. You can help prevent these accidents.

Using black construction paper, you can make cut-out falcons, or you can buy silhouettes of falcons, and tape them in the larger windows of your home or school, where light is often reflected. The cut-outs work like a charm, warning small birds to stay away.

Think About It

Think of some other ways to help reduce the problem of birds at airports so that live falcons do not have to be used.

Draw the grid on a piece of white paper, making the squares ten times larger than the squares in the illustration. Copy the outline of the falcon square by square. Use the outline to cut a model from black construction paper.

Surviving in the Environment

You have seen that organisms have similar needs but often very different ways of meeting them. Here are some other kinds of structural adaptations that help organisms to survive in their environment.

It's How They Look that Counts

For some organisms, appearance is very important. The way they look in their surroundings is essential to their survival. Take a look at some examples shown here to see just why.

Many plants need insects to pollinate them. Some, like the rose, lure insects with bright colours and attractive fragrances. Other plants have markings called "honey guides" that attract insects to them. Large dots near the edge of the foxglove's flower become more plentiful and more dense toward the centre where the pollen and nectar are.

This walking stick insect imitates the real twig. The walking stick's behaviour often fools birds and other predators that eat insects.

"Undercover"

Problem

Can your group make an imaginary animal that blends in with its surroundings and is not easily seen?

Materials

potato that has been painted with white paint
toothpicks
popsicle sticks
modelling clay
cotton wool
glue
tape
tempera paint
paint brushes

Procedure

1. First pick a particular place, such as a lawn, a wooded area, or a gravel path, where you will "hide" your animal.
2. Plan how you will use the potato and other materials to design your animal so that it will blend into the surroundings you have chosen. THIS IS A REAL CHALLENGE BECAUSE THE MAIN BODY PART WILL BE WHITE.
3. When you have finished making your animal, have a group member put it in the place the group has chosen and return to a class gathering point.
4. You will hunt for the animals other groups have hidden, while other groups hunt for your animal. (If you see your own, don't let anyone know!)

A tiger remains motionless in the surrounding plant growth as it stalks its prey. The tones and patterns of the tiger's coat blend with the plants, making the tiger almost invisible.

Finding Out

1. Which animals were easiest to find? most difficult?
2. Was your animal easily found? Analyse why or why not.
3. Evaluate your design and suggest ways it might be improved.

Finding Out More

4. What behavioural adaptations could you give your animal to improve its chances of survival?

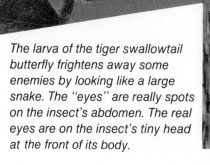

The larva of the tiger swallowtail butterfly frightens away some enemies by looking like a large snake. The "eyes" are really spots on the insect's abdomen. The real eyes are on the insect's tiny head at the front of its body.

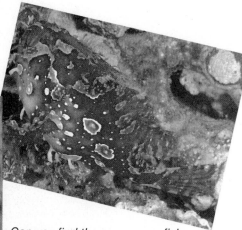

Can you find the sargassum fish hiding in the seaweed? If you were a tiny fish or crab and didn't sense the sargassum fish, chances are you would end up as a tasty morsel for it.

Coping with Hot and Cold

Birds and mammals are **warm-blooded**. Warm-blooded animals have internal receptors that control their body temperature, which is kept within a fairly narrow range. Your body temperature, for example, is usually fairly close to 37°C. If your temperature rises above that level, it usually means that you're not well.

All animals other than birds and mammals are **cold-blooded**. Cold-blooded animals have no internal receptors to control their temperature. They simply take on the temperature of their environment. When it is cool, you'll sometimes see cold-blooded animals such as dragonflies, turtles, or alligators basking in the sun. They spread themselves as flat as they can so that the sun warms as much of their body's surface area as possible. Cold-blooded animals only become active when their bodies reach a certain temperature.

In fact, the terms "cold-blooded" and "warm-blooded" don't really describe the situation well. A cold-blooded animal can be just as warm or warmer inside than a warm-blooded animal. On hot days, cold-blooded animals may even avoid the sun to stop themselves becoming too hot. Many move into burrows in the cooler earth or seek water or shade to keep cool.

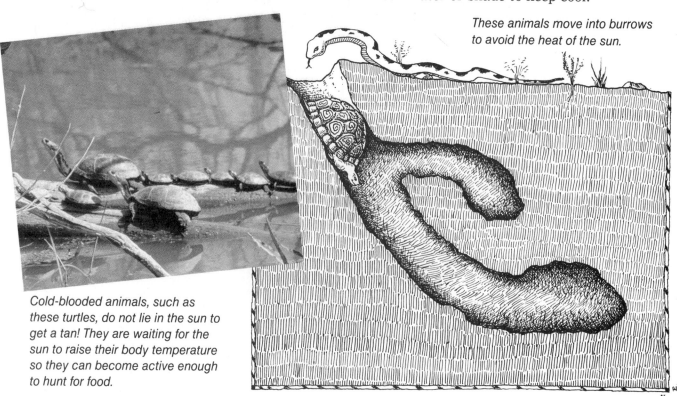

These animals move into burrows to avoid the heat of the sun.

Cold-blooded animals, such as these turtles, do not lie in the sun to get a tan! They are waiting for the sun to raise their body temperature so they can become active enough to hunt for food.

Being warm-blooded certainly has its advantages. For example, you don't have to wait for the sun to warm you up before you become active, the way a turtle, snake, or housefly does. On the other hand, a warm-blooded animal has to eat ten times as much food in a year as a cold-blooded animal of the same size. Which would you rather be, warm- or cold-blooded? Explain why.

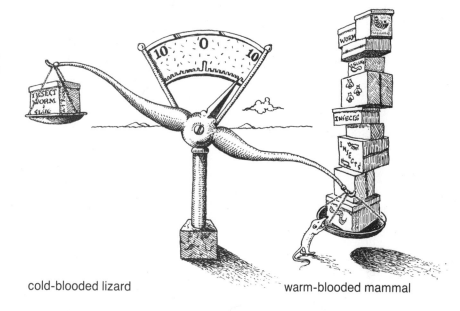

cold-blooded lizard   warm-blooded mammal

When Winter Comes

As you have seen, some organisms migrate when changes in their environment do not provide them with their needs. They simply leave one area for another. Other organisms have adaptations that allow them to survive during the period when the environment cannot meet their needs.

Many organisms are adapted to live for just one year. For example some plants are **annuals**; that is, they die at the end of a season, and leave only their seeds. New plants grow from these seeds the next year. Many of the insects that live in Canada are active only during warm seasons; some produce eggs that remain **dormant** (without growth) until the next warm season. Only a few, like the monarch butterfly, migrate.

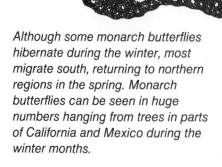

Although some monarch butterflies hibernate during the winter, most migrate south, returning to northern regions in the spring. Monarch butterflies can be seen in huge numbers hanging from trees in parts of California and Mexico during the winter months.

Winter Sleep

Some organisms go into a deep sleep, or **hibernate**, in response to winter's changes. When an organism hibernates, all its body processes slow down. As a result the organism uses very little energy. It needs no food, since its body's fat supplies it with all the energy it needs.

Activity 1-20

Hibernating Squirrels

John Li was very interested in knowing more about hibernating animals. He predicted the heart rate and body temperature of the thirteen-lined ground squirrel would be drastically slowed when the squirrel hibernated in the fall. He predicted also that these body processes would remain slow until the animal emerged in the spring. John measured the changes in heart rate and body temperature of the ground squirrel over a year. Here are some of his results in graph form.

Finding Out

1. Why do you think John made the predictions he did?
2. How do the graphs support his predictions?
3. Give the following information:
   (a) the heart rate of the active squirrel
   (b) the heart rate of the hibernating squirrel
   (c) the body temperature of the active squirrel
   (d) the body temperature of the hibernating squirrel.

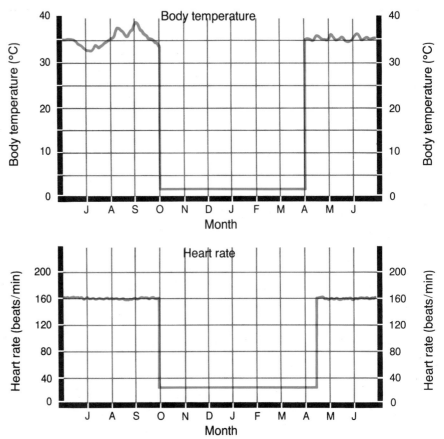

Heart rate and body temperature of a ground squirrel.

Some insects warm themselves for action by exercising, much as athletes do.

In order to keep their larvae at the right temperature for growth, ants move the larvae from one part of their nest to another depending on the temperature of that particular part.

When a beehive becomes too cool, worker bees can raise the temperature by becoming very active.

Humans keep their body temperature at 37°C by shivering and keeping active when they are cold, or sweating when they get too hot.

Truth Is Stranger than Fiction

Truth is stranger even than science fiction. Just take a look at some of these curious adaptations.

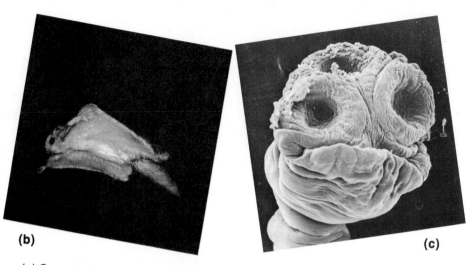

(a)

(b)

(c)

(d)

(a) Got you! The net-catching spider weaves a small sticky net and throws it over any insect that comes near!

(b) Do squirrels fly? No, but this flying squirrel is gliding from branch to branch. It certainly beats walking!

(c) This isn't the hero of a new science fiction movie. It's a tapeworm, only 1 cm long, that lives attached to a catfish's intestine. The "eyes" are really suckers. The tapeworm absorbs food from the fish through these suckers.

(d) This isn't a Hallowe'en mask. It's the blossom of a butterfly orchid. To bees and butterflies, it looks like an enemy and sometimes they attack it by striking it. The flower has "fooled" the insect into pollinating it.

Activity 1-21

Invent a Plant

Materials

almost anything (some suggestions: cellophane, modelling clay, crepe paper, wire, scissors, floral tape)

Construct an imaginary plant (or one modelled on a real plant) that is adapted to survive under certain environmental conditions. Choose one of the following plants.

- a plant that is lawn mower proof;
- a plant that can catch insects;
- a plant that can store water during dry periods;
- a plant that grazing cattle won't eat;
- a plant that can withstand high winds;
- a plant that can live on the surface of ponds or sloughs; or
- a plant with an adaptation you decide for yourself!

Be prepared to explain your design.

Finding Out

1. Evaluate either your own design or that of a classmate.
2. Give ideas on improving the design you have evaluated.

Working with Animals

Do you like animals? If so, you might like to become a veterinarian or a wildlife biologist.

At moments like this Jacqueline is certain that becoming a veterinarian was the right choice for her. Jacqueline had always enjoyed working with animals of all kinds, and her summer jobs included working as a dog groomer, a veterinary assistant, and a horse handler. Now, after completing five years of study in veterinary medicine, Jacqueline is looking forward to making her career out of what she likes best—caring for animals.

Owen Williams is a wildlife biologist. Part of his work is to help inform the public about wildlife, as he is doing here. However, he spends most of his time working on specific projects: How many deer are *too* many in a specific area? What kinds of regulations should we have for hunting and trapping? Is the wildlife in a certain area threatened? These are the kinds of questions that he and the people he works with try to answer in order to protect wild animals and the areas where they live.

Owen says, "At present, if you want to be a wildlife biologist, you should have at least a four-year university degree in the natural sciences. In many positions, it helps to have an extra year or two of specialized study after graduating. It's useful to have related work experience, such as summer jobs working with animals or as a junior forester. This may sound like a lot of training, but it is all extremely interesting. I'd recommend a career as a wildlife biologist to any student who is serious about working with living things in their natural environment."

1. Use the definitions to find the correct words for the word puzzle. When you have completed the puzzle, the highlighted letters will spell a word that describes the type of behaviour animals are born with.

(a) ■ ■ □ ■ ■ ■ ■ ■
(b) ■ ■ ■ ■ ■ □ ■ ■
(c) ■ ■ ■ ■ ■ ■ □
(d) ■ ■ ■ ■ ■ ■ ■ □ ■
(e) ■ □ ■ ■ ■ ■ ■
(f) ■ ■ ■ ■ ■ □ ■ ■
(g) ■ ■ □ ■ ■ ■ ■
(h) ■ ■ ■ ■ □ ■ ■ ■ ■ ■ ■ ■

(a) something in an organism's environment that causes it to react
(b) an organism's reaction to a change in its environment
(c) plants that live only one season, then die
(d) to go into a deep sleep with all body processes slowing down
(e) to move from one area to another according to the seasons
(f) a type of behaviour that results from the organism doing something more than once, or copying the behaviour of another organism
(g) a special body part that receives a certain type of stimulus
(h) how long it takes for an organism to respond to a stimulus (two words)

2. During the winter months, some organisms either hibernate or migrate.
(a) Explain what these two words mean.
(b) Give examples of an organism that hibernates and of one that migrates to survive the Canadian winter.

3. Give examples of a positive response and a negative response
(a) in a plant;
(b) in an animal.

4. Think of a common classroom situation where you would use at least three different senses to complete a task. Identify the senses you would use, and explain how they would help you with the task.

5. Explain the difference between instinctive behaviour and learned behaviour.

6. In Topic Six you were asked to design an animal and a plant to survive in a certain environment.
(a) Describe either the plant or the animal you designed.
(b) Explain what environmental conditions you had to consider to make your design.
(c) Identify the structural adaptations you used in your design.
(d) Describe any behavioural adaptations you might give your designed organism to improve its chances of survival.

> The living things in the photographs on page 2 are:
> (c) coral
> (f) barnacle
> (i) hydrozoan (a tiny sea animal)
> (j) stick insect

Unit Focus

- All living things grow, move, reproduce, respond to stimuli, produce or take in food, are made up of cells, and have a special chemical make-up.
- As some organisms grow, they undergo many structural changes.
- There are variations among members of the same species. Members of different species differ more from each other than do members of the same species.
- All organisms reproduce.
- Organisms have structural adaptations that help them survive in their environment.
- Movement in organisms may involve a change in shape, position, or location.
- Living things receive stimuli through their senses using receptors. Organisms respond to the stimuli in ways that help them to survive.
- An organism's life may depend on how quickly it responds to a stimulus.
- The ways an organism responds to its environment, or its behavioural adaptations, help it to survive.

Backtrack

1. Solve the crossword puzzle by using words from this unit.

Across Clues

3. reproduction, growth, and adult life (two words)
5. special features living things have
7. a difference between organisms of the same species
10. an organism's pattern of responses
11. a vital sign which indicates the beating of the heart muscle
13. alive, but not active; in a state of no growth
15. changing shape, position, or location
17. a _____ thing can move, grow, take in food, react, and reproduce
18. to be in a state of very deep sleep
19. behaviour that an organism is born with

Down Clues

1. group of organisms that are similar in size and shape
2. the number of times you inhale and exhale every minute (two words)
4. living things
6. sight, touch, or smell
8. a body structure or a behaviour
9. the process of organisms producing offspring similar to themselves
12. a pattern that shows at what speed an organism increases in size (two words)
14. a method of plant reproduction where a stem or a leaf is placed in water or moist sand
16. an organism may do this by responding to the same stimulus several times

2. What are the main characteristics of life that all organisms have in common? (Try to name seven.)

3. Explain the difference between the following terms:
   (a) movement and locomotion
   (b) stimulus and response
   (c) hibernate and migrate
   (d) structural adaptation and behavioural adaptation
   (e) instinctive behaviour and learned behaviour

4. Should a leaf that has fallen to the ground in the autumn be classified as a living thing or a non-living thing? Explain your answer.

5. Explain how your thumb is a structural adaptation.

6. You are hungry. A friend says you may have an apple, but only if you make the right choice. You are blindfolded and given two apples. One is real, the other is artificial.
   (a) What senses must you rely on to help you make the right choice?
   (b) Assuming you have made the right choice, list the structural adaptations that allow you to eat the apple.

7. Decide which term does not belong in each group. Write the term in your notebook and then tell why it does not belong.
   (a) an earthworm's skin, plant stomata, a dog's skin, a fish's gills
   (b) a twig caterpillar, a cat, a fawn, a chameleon
   (c) growth rate, breathing rate, pulse rate, body temperature

Synthesizer

8. Oxygen enters and carbon dioxide passes out of an organism's body. Explain why this process is necessary to living things.

9. Plants such as the rose and the foxglove are adapted to attract insects.
   (a) Why is this important to the plant?
   (b) What would happen if the insects were not attracted to the plant?
   (c) Do the insects benefit in any way from the plant?

10. Living things are made up of cells. Are all cells the same shape? Explain your answer, or illustrate it in some way.

11. To which characteristic of life does the statement "life produces life" refer?

12. Lions and dandelions both need food to live.
    (a) Explain the methods each species uses to obtain its food.
    (b) List a structural adaptation of each that assists it in obtaining its food.

13. Living things adapt to seasonal changes in their environment in a variety of ways. Identify one way in which each of the following organisms is adapted to survive in winter.
    (a) a squirrel
    (b) a pet cat
    (c) an American robin
    (d) a black bear
    (e) a tulip
    (f) an earthworm

14. Jason told Maria that he had seen a hawk perched at the top of a poplar tree on a very still day and that both organisms were doing absolutely nothing. Maria said that she didn't agree and that there was lots of activity going on. With whom do you agree? Give several reasons for your decision.

15. Adaptations allow organisms to continue to survive in their environments. For instance, our teeth are structural adaptations that aid us in taking in our food. Describe an adaptation that helps the organism survive in each of the following situations:
    (a) turnip—structural adaptation for obtaining water
    (b) dog—behavioural adaptation for responding to danger
    (c) duck—behavioural adaptation for protecting its young
    (d) strawberry—structural adaptation for reproducing
    (e) human—behavioural adaptation for gaining attention
    (f) cactus—structural adaptation for drought conditions
    (g) polar bear—structural adaptation for Arctic waters
    (h) meadowlark—behavioural adaptation for reproduction
    (i) beaver—structural adaptation for balanced movement in water
    (j) painted turtle—behavioural adaptation for controlling body temperature

16. If you toss a ball of paper into the wastebasket, you are responding to a stimulus (seeing the wastebasket). The structural adaptations that allow you to do this task include: hands and fingers for grasping, muscles and joints for grasping and tossing, eyes for seeing the target, and the brain for co-ordinating the action. List at least three adaptations that allow you to perform each of the following tasks (more than three if you can!):
(a) getting out of bed after the alarm goes off
(b) bandaging a cut finger so it will heal

17. Making predictions and designing activities to test these predictions are some of the skills that scientists use to study things that interest them. Look back at how Niko Tinbergen found out more about the digger wasp's behaviour. Analyse the steps he took by answering the following questions:
(a) What was the problem he wanted to investigate?
(b) What was his prediction about the wasp's behaviour?
(c) What were the steps in his experimental procedure?
(d) What were his observations based on the steps he took?
(e) How did Dr. Tinbergen interpret his data?

18. Describe how a particular mechanism, made possible by modern technology, allows people to function more easily in difficult situations. Consider the needs of either (a) or (b).
(a) a sense-impaired person
(b) a long-distance space traveller

19. Many kinds of hawks eat young ducklings.
To a duckling on the ground, a hawk above looks like (a).

(a)

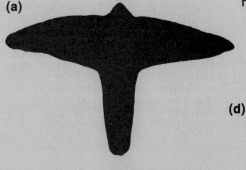

A duck flying overhead looks like (b).

(b)

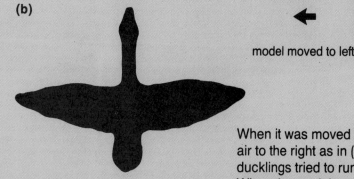

A scientist used a model like the one shown to see how young ducklings would respond to it.

(c)

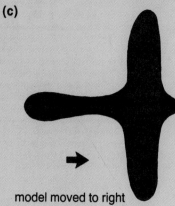

model moved to right

(d)

model moved to left

When it was moved through the air to the right as in (c), the ducklings tried to run or fly away. When the model was moved through the air to the left as in (d), they did not respond. Explain why you think the ducklings responded differently to the model, depending on the direction it was moved.

Structures and Design

Have you ever sat on a chair that collapsed beneath you, or seen bookshelves that sag in the middle? You take for granted the structures you use every day—your body, a chair, your school building—until they give way!

In this Unit you will be looking closely at structures, and at the way they are put together. You will compare structures made by people, such as buildings, bridges, and furniture, with structures in the natural world, such as shells, bones, and twigs. You will discover how to choose the best materials for making structures and why one design is stronger than another.

You can learn about the importance of shapes by designing, building, and testing your own structures. Learning about structures will help you discover why tables and chairs, bridges, and stadiums are designed and built the way they are. People who plan and build large structures must think about the cost, safety, and appearance of the structures, as well as their design.

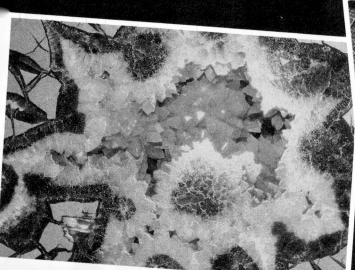

Discovering Designs

A soaring hang glider looks like some kind of strange, giant bird. Both the hang glider and the bird are **structures** used for flying. A structure is an object built up from one or more parts. The parts in the hang glider and in the bird include a body and wings.

Look at the shape of the wings—they are broad and flat. Look at how the wings are arranged—they stretch out on either side of the body. Both the hang glider and the bird therefore have a similar **design**. A design is the shape of the parts and the way they are put together in a structure.

The bird is a **natural** structure. The hang glider is a **manufactured** structure, or one made by people. There are many other examples of similar designs in natural and manufactured structures.

Shapes in Design

Some designs consist of a single basic shape repeated over and over. This type of design is found in natural structures such as honeycombs, and in manufactured structures such as brick walls. One advantage of this type of design is that one or just a few repeating shapes can be put together in hundreds or thousands of different ways. In manufactured structures, this can save a lot of money. A brick factory, for example, can supply the same shape of bricks to build structures as different as a garden path, a house, or a well.

Probing

1. (a) What shape is each brick?
   (b) How many sides does each brick have?
2. Why do you think each brick in a wall is not fitted exactly on top of a brick in the layer below it?
3. How many sides does each of the cells in the honeycomb have?

Matching Designs

Problem

What designs do natural objects and manufactured objects have in common?

Procedure

1. In your notebook, make a table similar to Table 1 to record your observations.

2. In the first column of your table, list all the natural structures in the illustrations.

3. Match up each natural structure with a manufactured structure that has a similar design. List the manufactured structures in the second column.

4. In the third column, describe what part or parts each pair of structures has in common.

Table 1

NATURAL STRUCTURE	MANUFACTURED STRUCTURE (MADE BY PEOPLE)	DESIGN IN COMMON

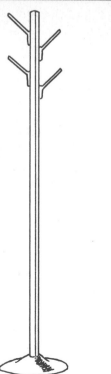

Finding Out

1. Do you think the designer of each manufactured structure knew about the natural structure it resembles? Explain why or why not.
2. For each pair of natural and manufactured structures, state whether you think they are used for exactly the same purpose. Explain why or why not.

Finding Out More

3. List three other pairs of natural structures and manufactured structures that have a similar design. How many examples can your class think of altogether?

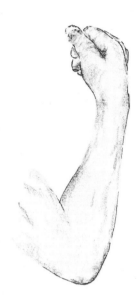

Tangram: an Ancient Chinese Puzzle

You may have seen this kind of puzzle before. The aim is to make different structures by using only the seven pieces shown here. An expert at the game can design hundreds of different structures showing animals, people, machines, buildings, and many other things.

1. Copy the tangram pieces onto a sheet of paper. Carefully cut along the lines so that you end up with seven pieces of paper.
2. Reassemble all your seven pieces to make first a duck, then a sailboat. The pictures at the bottom of the page here will help you.

3. Now see if you can make these two figures: a kangaroo and a watering can.
4. Design four more structures of your own, some natural and some manufactured. Draw the outline of each structure. Then exchange your outlines with a partner. Assemble the pieces to fit your partner's outlines.

Separating the Parts

You have seen that the repeated use of a few basic shapes can produce many different structures. But most structures are not designed in this way. Instead, they are made up of parts that look very different from one another.

For example, a bicycle is made up of several major parts, such as the frame, handlebars, saddle, and wheels. These major parts are themselves made up of smaller parts—the wheel consists of the rim, spokes, axle, inner tube, and tire.

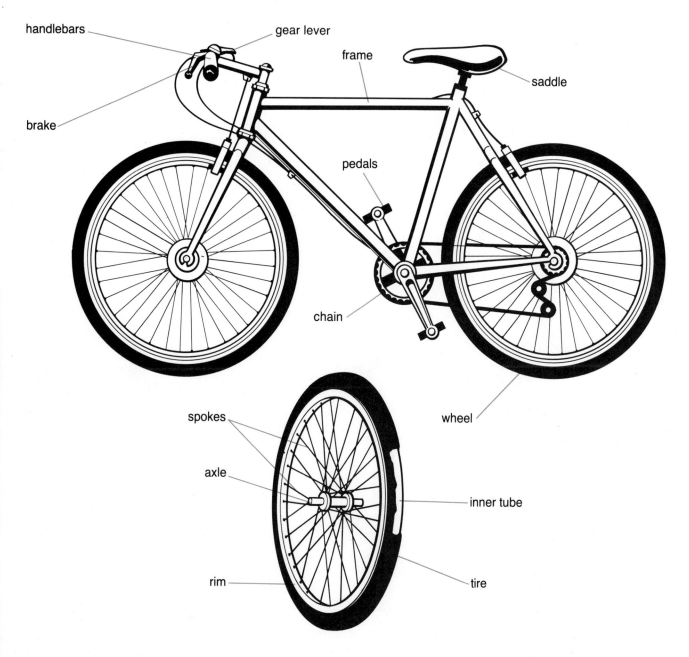

Breaking It Down

Problem

How many different parts can you find in manufactured structures?

Materials

structures such as a chair, desk, skateboard, framed picture

Procedure

1. In your notebook, make a table similar to Table 2 to record your observations. One example has been done for you to help you get started.
2. Examine each manufactured structure.
   (a) List the major parts that make up each structure.
   (b) Describe the shape and quantity of each part in the structure.

Finding Out

1. (a) Which shapes were most common?
   (b) Which shapes were least common?
2. Were structures most often made of single, unlike parts, or did they have several similar parts?
3. Name any of the major parts that could be further divided into smaller parts.

Finding Out More

4. (a) Could any of the structures you examined be made with fewer parts and still be used for the same purpose? For example, could a chair be made with three legs, two legs, or one leg? Why or why not?
   (b) Could the parts of each structure have a different shape? Why or why not?

Table 2

STRUCTURE	PART	SHAPE	QUANTITY
chair	seat	circle	1
	back	rectangle	1
	legs	cylinder	4

5. How many structures can you find in the cartoon on page 83 that look similar to one another?

6. How many parts can you find that look similar to one another?

7. Are these similar-looking structures and parts designed for the same purpose? Explain.

8. (a) Which actions would not be safe in real life? Which structures would not be safe? (b) Describe how you would change the design of one of these structures to make it function better.

Understanding Design

Understanding designs can help you to see the world in a different way. You can look at any structure—natural or manufactured—and imagine how to take it apart and put it back together again. You can also begin to see how one structure, or a part of it, may be similar to another structure or part. Look, for example, at the illustration shown here. You see ropes, wheels, wings, and other parts used in many different ways. Identify as many as you can.

In the rest of this Unit, you will see how recognizing such similarities (and differences, too!) can be useful in designing the many devices and buildings we use each day.

Design and Function

Very early in human history, people realized that they could use some parts of animal and plant structures to carry out certain **functions** (purposes). At first, people used these natural structures exactly as they found them. The shape of each structure suggested how it could be used. For example, a tortoise's shell or a gourd became a scoop or a bowl, and a sharp tree thorn became a needle.

The first people to see that a coconut shell would make a useful container were using thought and imagination. In Activity 2-4, use your own imagination on some common objects around the classroom.

Can you name other examples of natural structures that are used by people? Is the function of these structures when people use them the same as their function in the natural world?

Humans are not the only
ones to find uses
for natural structures.

Beavers use logs, twigs,
and mud to build a dam.

A robin uses mud, grass, and twigs to
build a nest.

A raccoon family
uses a hole
in a tree
for shelter.

The longtailed shrike is using a thorn to
store its next meal: a dead frog.

The chimpanzee is about to use this long stick as a ladder
to get food that is too high for it to reach.

You Use It for *What?!*

This is a thought activity. Pretend that you have never seen some common structures before, and that you don't know their function. Using only your observations about the design of the structures, suggest some of the ways in which they *could* be used.

1. Look at the jar and the ruler. Remember that you have never seen a jar or a ruler, so you don't know what they are used for. From your observations of the appearance of each structure, decide how it could be used. For example, the jar could be used as a small greenhouse, and the ruler could be used as a stir-stick.

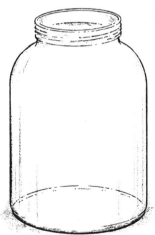

2. Now work with a partner. Take turns showing each other simple structures around the room. The partner who is "guessing" must describe as many functions as possible for that structure in about one minute. You can be funny, but the uses you suggest must make sense.
3. Record the name of each structure and your list of functions.

Finding Out

1. (a) For which structures was it easy to imagine different functions?
(b) For which structures was it difficult to imagine different functions? Why?
2. Were there any structures that could be used for only *one* function—the function they were designed for and nothing else? List these structures.

Extension

3. (a) Next time you visit a museum, pioneer village, or other historical site, look for objects that were once commonly used but are now seen only in museums. Try to figure out the function of some of these objects without asking someone or reading the labels that explain them.
(b) What did you look for in each object to help you figure out its function?
4. Find pictures of unfamiliar manufactured structures in a library book. They could be structures that were only used long ago, or ones that are used by people in other cultures. Show the pictures to your friends or family as a puzzle, and have them guess what the structures might be used for.

What useful function does the log serve?

From Structures to Construction

Some simple objects, such as rocks, shells, and sticks, can be used for a variety of functions. But there is a limit to the usefulness of a structure that consists of only one part. Early humans soon discovered that they could change the shape of natural objects, or fit objects together. In doing this, they designed structures for particular purposes. For example, they sharpened stones to make blades. They fastened stone blades onto wooden sticks to make spears. The process of shaping parts and of fitting different parts together was the beginning of construction and manufacturing—the beginning of **technology**.

In what way is the seashell limited in its usefulness as a container? How is the basket an improvement over the seashell?

In the following Activity, you are asked to use a number of simple objects to make structures for carrying out different tasks. Think about how you might use each object. How can you change its shape, or combine it with other objects? Plan your designs carefully, and alter them if they don't work out at first.

By Design

Problem

How can you use the same few objects to make structures that have different functions?

Materials

2 drinking straws
5 metal paperclips
2 small containers such as
  175-mL plastic yogurt
  containers
250 mL of fine sand
cardboard tube from centre of
  toilet roll or paper towelling
small rubber ball or marble
cardboard (about
  20 cm × 20 cm)
scissors
2 large rubber bands
ruler
large ball of modelling clay
adhesive tape
water

Procedure

1. There are eight tasks in this Activity. Your goal is to carry out as many of these tasks as you can.
   • For each task, design structures using the smallest number of objects you can.
   • For each task, the function of the structure you are asked to design is printed in *italic (slanted)* type.
   • You choose which of the objects to use for the structure in each task.
   • You may use the same objects in different ways for different tasks.

   • You may modify the objects if you need to. (For example, you can cut the cardboard square into other shapes if you wish.)
2. (a) In your notebook, make a table with four columns. The heading for the first column is *Tasks*; for the second column *Materials used*; for the third column *Design*; and for the fourth column *Observations*.
   (b) Complete the columns as you carry out the tasks.
   (c) Use the fourth column to sketch or describe your structures. Record any problems with your design.
3. Some tasks are more challenging than others. But don't quit a task after only one try. A different design for the structure may be all that's needed. If, after some trial and error, you can't carry out a task, go on to the next one. (Note: Most of the tasks can be done in more than one way.)
4. Read the task through and make sure you understand what you are to do. Use this method to help you as you work through the tasks.
   • *Plan* how you will use the objects to carry out the task.
   • *Design* the structure.
   • *Test* your structure.
   • *Evaluate* your structure: Did it perform the task? If not, how could you redesign the structure?

TASK 1

• Design a structure that you can use to *contain* about 1 mL of water.

TASK 2

• Place a marble or ball in the bottom of an empty container.
• Design a structure that will *shelter* the ball, so that you can pour sand into the container without the sand touching the ball.

TASK 3

• Design a structure that you can use to *support* a container of water about 10 cm above the surface of the bench. (Work near a sink so that if your structure fails, clean-up will be easy!)

TASK 4

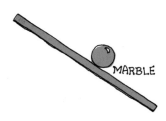

- Cut a piece of cardboard approximately 20 cm × 6 cm.
- Place one end of the cardboard on a container or other suitable object to make a ramp.
- Design a structure that will *support* a marble or small ball about halfway up the ramp without the ball rolling down.

TASK 5

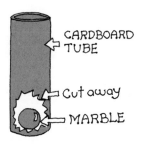

- Stand a cardboard tube upright on the bench.
- Drop a marble or small ball into the tube.
- Design a structure that you can use to *lift* the ball from the tube without moving the tube.

TASK 6

- Roll some modelling clay into a ball.
- Design a structure that you can use to *separate* the ball of clay into three parts.

TASK 7

- Half fill a container with sand.
- Place a second container about 15 cm from the first.
- Design a structure that you can use to *transport* some of the sand from the first container to the second. Explain how you would use it.

TASK 8

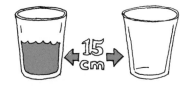

- Half fill a container with water.
- Place a second container about 15 cm from the first.
- Design a structure to *transport* some of the water from the first container to the second. (Work near a sink so it will be easy to clean up any spills.) Explain how you would use it.

Finding Out

1. (a) Which tasks did you find easiest?
   (b) Which were more difficult? Why?
2. (a) Which materials did you use most?
   (b) Which materials did you use least? Why?
3. How do your answers to Questions 1 and 2 compare with those of others in your class?
4. Did you change the design of any structure after you had tried and failed to solve a task? If so:
   (a) what was the task?
   (b) what part of your structure did not perform in the way you expected, and why?
   (c) did your new structure require:
   • a change in the shape of the parts?
   • a change in the materials used?

Extension

5. In this Activity, you were asked to design structures using the *least* number of parts. Think about the tasks again, then design a structure for one task using as many parts as you want. Test your design and then evaluate it.
   (a) Does using more parts make it easier to design a structure?
   (b) Is the structure that uses more parts better for carrying out the function than the first structure you built? Explain.

Classifying Functions

In today's world, you are surrounded by thousands of complex structures with many different parts. But although there are many more manufactured structures now than in the past, there has not been a large increase in the number of functions they perform. For example, a canoe, helicopter, train, bicycle, and automobile are structures that have very different designs from one another, but they are all used for the same function: transportation.

Similarly, a tent, an igloo, a castle, and a modern house are all built to provide shelter. Transportation and shelter are two of the most important functions for which people build structures. In Activity 2-6, you will look at some other functions.

What's the Purpose?

Procedure

1. In your notebook, make a copy of Table 4 to record your observations.
2. Select six structures from the illustration below and list them in the first column of your table.
3. Choose the function from Table 3 that you think best describes the purpose for which each structure was designed. Write the function in the second column of your table.

The designs of these structures are obviously very different. But the functions are still the same.

Table 3 *Common Functions of Manufactured Structures*

grasping
marking
supporting
containing
lifting
fastening
breaking
protecting
sheltering
transporting
separating
communicating

4. If any structures have more than one function, write the other functions in the third column of your table. (See the example given.)

Finding Out

1. Do structures that have the same function always have a similar design? Explain why or why not.
2. Do you see any connection between design and function? Explain.

Finding Out More

3. (a) Choose one function from Table 3. Name five structures that are designed and built to carry out this function.
   (b) How does the design of each structure help it carry out this function?
4. (a) List the various kinds of buildings in your community, such as stores, houses, factories, and churches. For each type of building, write the main function for which it is used.
   (b) Can you see any connection between the design of a building and its function? Explain.

Table 4

STRUCTURE	MAIN FUNCTION	OTHER FUNCTIONS
pen	marking	communicating

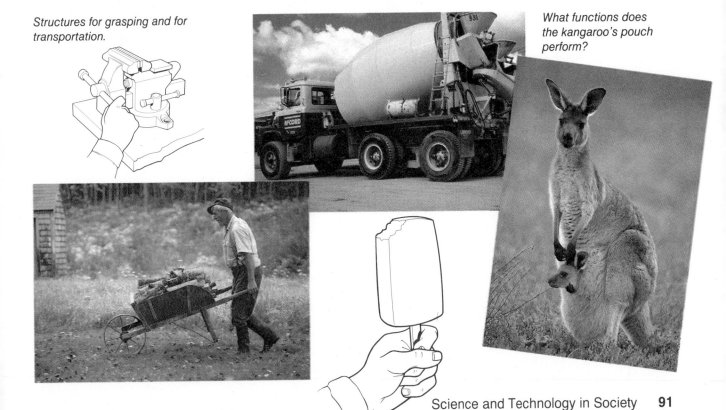

Structures for grasping and for transportation.

What functions does the kangaroo's pouch perform?

1. In your notebook, complete the following sentences:
   (a) An assembly of parts is a ■□ ■■■■■□■■■ .
   (b) The way the parts are put together is its ■■■■■■□ .
   (c) Similar ones can be found in both ■■□■■■■■■■□■ objects and the ■■■■■■□ world.

2. The highlighted letters spell out an example of what is defined in sentence (a) above. Use your answer in sentence (c) to give two actual examples of this item.

3. Draw Table A in your notebook. Match the structures in the first column with the parts in the second column. In the third column, write how many of each part are found in each structure.

Table A

STRUCTURE	PART	NUMBER OF THAT PART
skateboard	page	
scissors	case	
book	handle	
cassette	cover	
	blade	
	spool	
	wheel	
	tape	
	platform	

Table B

STRUCTURES	FUNCTION OF EACH	HOW ARE THEY THE SAME?	HOW ARE THEY DIFFERENT?
(a) postcard telephone	communicating	both send messages	written spoken

4. Fill in a table similar to Table B for each pair in the list.
   (a) postcard, telephone
   (b) bicycle, motorbike
   (c) bottle of pop, bottle of shampoo
   (d) elevator, escalator
   (e) stairs, escalator
   (f) curtain, bedsheet
   (g) curtain, venetian blinds

5. "A chair is designed for one function but can be used for several others." Explain this statement.

6. In your notebook, sketch a design for a structure that has at least three functions: lifting, transporting, and supporting. The structure you design can do even more if you choose to have it do so.

7. Some items used everyday are shown here.
   (a) Make a list of all the items that can be used for containing.
   (b) Make a second list of all the items that can be used for digging.
   (c) Could any items serve both functions?
   (d) List two natural objects that could be used for each of these functions.

Materials Matter

It is not enough for a structure to have a good design. For a structure to carry out its functions properly, it must also be made of suitable **materials**. A rubber hammer, for example, has the right design but the wrong materials. In order to carry out its function of driving in nails, a hammer must be made of things that are hard, such as wood and steel.

Before you can choose the best materials for a particular structure, therefore, you have to consider the **properties** of those materials. Properties include such things as strength, stretchability, and hardness. Most structures are made of several materials, each of which has different properties. For example, think about a pencil. It is made of only two materials —wood and graphite (also called black lead). The wood must be firm and strong to protect the lead and keep the pencil from breaking easily. But it must not be so hard that you cannot sharpen the pencil. The lead must be soft, so that it produces marks when drawn across a piece of paper. The properties of both materials together allow the pencil to function.

Choosing the Best Materials

Choosing the best materials for a structure isn't always easy. Designers sometimes find that one property conflicts with another. For example, a structure built from lightweight metal, such as aluminum, has the advantage of being easy to move. But aluminum may not be strong enough to fulfill the structure's function. Most materials have some disadvantages. For example, steel, which is widely used in bridges, cars, and buildings, has the disadvantage of rusting.

Sometimes the best material—even if it has all the right properties—may not look attractive. For example, aluminum siding is an ideal material for protecting the outside walls of a house from wind and rain. Some people, however, prefer wooden siding, even though they have to paint it. Sometimes the best material may be too costly to use in a certain structure.

Wrapping up Liquids

For this Activity, look in your kitchen cupboards at home or on the shelves of a local store, and make a list of 5 different liquids. Beside the name of each liquid, write:

(a) the type of container used to hold the liquid;

(b) the material used to make each container;

(c) the size and shape of each container.

Using your list and the picture shown here, answer the Finding Out questions.

Finding Out

1. How many different packaging materials did you find?
2. What are the properties of each material?
3. What is the advantage of each property? (For example, a property of glass and clear plastic is that you can see through the material. The advantage of this property is that you can see what the liquid looks like).
4. Are similar kinds of liquids packaged in more than one type of container? If so, why?
5. (a) What properties of the package do you think are important to the people who make and sell the package? (b) What properties do you think are important for consumers (the people who buy the product)?

Finding Out More

6. What kind of milk package does your family usually buy? List one advantage and one disadvantage of this kind of package.
7. What kind of milk can you buy in cans? Why do you think this milk is sold in cans and other kinds of milk are not? (If you are not sure, you will find out the answer in Unit Five on micro-organisms.)
8. List some materials that are never used for packaging liquids. What properties make them unsuitable for this purpose?

Strength of Materials

Some parts of a structure are often under **stress**. For example, your shoelaces are under stress when you tie them and all the time that you are wearing your shoes. Bookshelves are under stress from the books they are supporting.

One of the most important properties of materials is their strength. Materials must be strong enough to withstand all the stresses that are placed on a structure during normal use.

Some materials are stronger than others. The strength of a material must be suited to the structure in which it is used. Paper, for example, is strong enough to be used to make envelopes and shopping bags. But, of course, paper isn't strong enough to be used to make automobiles or airplanes! You can compare the strengths of different materials by looking at the way they respond to a **load**. For example, a simple test shows that a block of wood can support a greater load than a marshmallow of the same size!

The type of stress exerted by the book (the load) in this example is called **compression**. Compression acts to squash a structure.

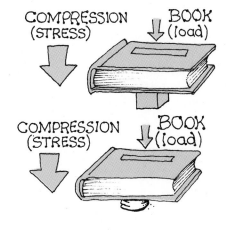

COMPRESSION (STRESS)   BOOK (load)

COMPRESSION (STRESS)   BOOK (load)

An essential material used in many buildings is mortar, a mixture of sand and cement that holds bricks together. As a brick wall is built higher, each layer of bricks adds to the compression on the bricks below it. The mortar, as well as the bricks, must be able to withstand the stress of compression without breaking. In Activity 2-8 you will put mortar under compression, and compare the **compressive strength** of samples of mortar made according to different recipes.

Squash It!

Problem

What is the compressive strength of different samples of mortar?

Materials

sand
cement
water
containers for mixing mortar, such as 175-mL yogurt containers or plastic cups
strips of non-absorbent paper (about 20 cm × 1 cm) for making moulds
stiff plastic sheet
C-clamp or vise (load)
safety glasses
safety apron
small scoop or plastic spoon

Procedure

1. Mortar is made by mixing cement and sand with water. Follow the instructions below for mixing three different samples of mortar.
2. Predict which sample of mortar will be strongest and which will be weakest. Give a reason for your prediction. (Hint: Think about the properties of a sandcastle.)

CAUTION: When you test the compressive strength of the mortar, make sure you wear safety glasses and a safety apron.

---

3. When your samples of mortar are dry, test their compressive strength by slowly crushing each one with a C-clamp or a vise. Record the number of times you turn the screw on the clamp before the sample breaks apart.

Finding Out

1. Were your predictions correct? If not, suggest a reason.
2. Rate the importance of each mortar ingredient to the strength of the mortar.

MAKING MOULDS

STIFF PLASTIC SHEET

CEMENT SAND

A
7 SPOONFULS OF SAND
1 SPOONFUL OF CEMENT

B
4 SPOONFULS OF SAND
4 SPOONFULS OF CEMENT

C
1 SPOONFUL OF SAND
7 SPOONFULS OF CEMENT

Use the three mixtures to make three samples of mortar. Add a little water gradually to each mixture to make a thick paste. Pour the mixtures into the moulds you have prepared.

Pushing Together

It is important to know the compressive strengths of different materials when building structures that must carry a load on top of them. A road, a chair, and a shelf are all examples of structures that must have a high compressive strength to carry out their normal functions.

Pulling Apart

You know that compression is a stress that acts to squash a material. There is another type of stress that acts to pull a material apart. This stress is called **tension**. For example, when you pull tightly on a rope, you put it under tension. The ropes or chain on a swing, a fishing line, and the steel cables on a crane are parts of structures that must have a high **tensile strength**.

Materials vary in their ability to resist the stress of tension. You can test and compare the tensile strength of materials by gradually pulling on them with increasing loads until the material breaks. In the next Activity you will compare different loads by measuring their mass. Read *Skillbuilder Three* at the end of the book to find out more about measuring mass.

The dog's leash is under tension. If it has a low tensile strength, it will break.

Pull It!

Problem

What is the tensile strength of some familiar materials?

Materials

30-cm lengths of different threads such as sewing thread, knitting yarn, string, dental floss
50-cm (or longer) wooden bar
standard masses or other small, heavy objects such as nails to act as load
bag or plastic container to hold the load
balance

Procedure

1. Support the wooden bar across the backs of two chairs or any other supports. The bar should be at least 50 cm above a surface.

2. Tie one sample of thread securely to the bar.
3. Attach the lower end of the thread to the bag. Gradually add standard masses to the bag until the thread breaks. Measure and record the total mass of the load.
4. Repeat this test of tensile strength for the other samples of thread. Before each test, predict what mass the sample will support. Then test your prediction.

Finding Out

1. List the thread samples in order of their tensile strength, with the strongest at the top and the weakest at the bottom.
2. What was the difference in mass between the greatest load and the smallest?

Extension

3. Design an experiment to measure the tensile strength of different samples of paper, such as newspaper, paper towelling, writing paper, grocery bag paper, and waxed paper. Make a prediction. List the samples from the strongest to the weakest. Then test your prediction.
4. Is a sample of paper equally strong when pulled from all directions? In other words, is it as strong if you pull it across the width of the sample as it is when you pull along the length? Design an experiment to test these tensile strengths.

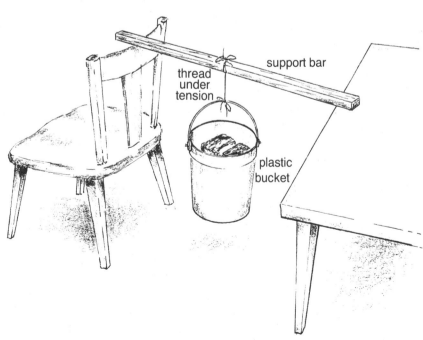

thread under tension

support bar

plastic bucket

Material Failure

You can see that materials must be strong enough to withstand the normal stresses on a structure, such as tension and compression. If the wrong material, or a poor quality material, is used in a building, for example, it could lead to disaster.

There is another factor that designers and engineers must consider. Sometimes the properties of a material change after the material has been in use for a length of time. A familiar example of this is what happens to the metal used to make the body of an automobile. After exposure to a few Canadian winters, the metal begins to rust. Rusting makes the metal weaker. Rusting is a major problem in materials that contain iron. For example, some bridges that contain iron have become dangerous and unusable after heavy rusting.

In the 1950s, people discovered another problem caused by changes in metals. Pressurized cabins in some aircraft failed during flight. Investigators found that these failures were caused by **metal fatigue**—a weakening of the metal caused by too much stress. Metal fatigue causes the metal to change so that it breaks under a much smaller load than it did when it was new.

In 1988 a large part of this airliner's fuselage was ripped away as the airplane was about to land in Hawaii. Investigators found that, over time, salt water had rusted parts of the fuselage, leading to metal fatigue.

New Materials

Scientists and engineers attacked the problem of metal fatigue in two ways. One way was to develop new metals that were resistant to fatigue. They found that a mixture of two or more metals provided the properties they needed. Metals made from combinations of other metals are called **alloys**. Alloys have properties that are different from the properties of the metals from which they are made. Table 5 shows some commonly used alloys.

Investigators also found that metal fatigue had occurred only at certain points on the aircraft. These were the points that had the heaviest stress. Another solution to metal fatigue, therefore, was to design aircraft in different ways. The new designs spread loads more evenly across the whole structure. This avoids the problem of having stress concentrated on only a few points. You will look at ways of designing for strength in Topic Four.

Table 5 *Some Desirable Properties and Uses of Alloys*

METAL COMBINATION OR ALLOY	PROPERTIES	USES
copper + nickel	• light • hard • resists corrosion (rusting)	• coins • propeller shaft of aircraft
copper + zinc (brass)	• shiny and attractive • strong • resists corrosion • can be polished	• ornaments • parts of electrical plugs • door fittings
copper + tin (bronze)	• resists corrosion • shiny and attractive • can be polished	• ship propellers • statues
iron + carbon (steel)	• can be rolled into thin sheets • strong	• car/truck bodies • construction
iron + carbon + chromium + nickel (stainless steel)	• no corrosion • shiny • strong • hard	• cutlery • sink tops • surgical instruments
aluminum + copper	• light • strong • resists corrosion • melts only at very high temperatures	• aircraft bodies

Other Properties that Matter: a Reminder

You have been thinking about the properties of materials that make them strong. Preventing the failure of a structure is one very good reason to select one material instead of another. But what other properties are important when deciding which material to use? When two materials both have the necessary strength for a function, the decision which to use may be based on *cost* and *appearance*. For example, steel and stainless steel are equally strong. Steel is inexpensive but not very attractive. Stainless steel is expensive but has an attractive appearance. Which is used to make cutlery? Which is used to make a truck? Cost and appearance are also considered when designers decide whether plastic, wood, or some other natural material such as marble should be used to make a certain structure. Keep these other properties in mind as you design your own structures later in this Unit. For some structures, the strength of the materials may not be your most important consideration.

The Interpretive Centre at Head-Smashed-In Buffalo Jump, a Unesco world heritage site near Fort McLeod, Alberta. The appearance of this museum is an important part of its design. The building was designed to blend in with the cliff where the buffalo used to be hunted.

Designing for Support and Strength

You have probably seen a bridge like this before. It is an example of a simple structure called a beam. A **beam** is a strip of material used horizontally to support a load. The man on the bridge is a load. He is putting a stress on the bridge. A slightly greater load may be just too much stress!

This Topic will help you understand structures that use beams. You will investigate the effects of loads on beams, and find out how to support and strengthen beams. As you see in the illustration, one effect of a load on a beam is to make the beam bend. In the next Activity, you will examine the stresses involved in bending.

The Bending Beam

Problem

Does bending under a load involve tension or compression?

Materials

several pleated straws
books to act as a support

Procedure

1. Take a new bendable straw and sketch the appearance of the pleated section in your notebook.
2. Hold the straw by the ends and gently pull on the ends. Sketch the appearance of the pleats.
3. Push gently on both ends and sketch the appearance of the pleats.
4. Make a simple bridge by placing the straw across two piles of books.
5. Push down with your finger near the pleated section. Sketch the appearance of the pleats.

Finding Out

1. Were the pleats under tension or compression
   (a) when you pulled on the ends?
   (b) when you pushed on the ends?
2. When you bent the straw bridge by applying a load with your finger
   (a) which part of the bridge was compressed?
   (b) which part was under tension?

Finding Out More

3. Suggest one way you could make the straw bridge support a greater load without bending.

Bending and Breaking

A beam may respond to a large load by bending. Bending involves *both* tension and compression. If the load is too great, however, the beam will break. It may break as a result of failure in either the tensile or the compressive strength of the beam, as you can see in the picture.

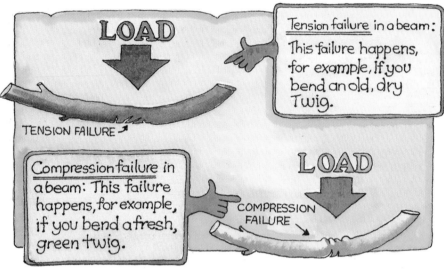

LOAD

Tension failure in a beam: This failure happens, for example, If you bend an old, dry Twig.

TENSION FAILURE

Compression failure in a beam: This failure happens, for example, if you bend a fresh, green twig.

LOAD

COMPRESSION FAILURE

A twig is a simple beam. You can see that bending involves both tension and compression. If either the tensile strength or the compressive strength fails, the beam will break.

The problem of bending or breaking can be avoided in three different ways.

1. You can make the beam from a stronger material.
2. You can add a part to support a beam.
3. You can change the design of the beam to make it stronger.

In Topic Three you looked at the strength of materials. Now look at the other two ways to avoid bending and breaking.

Designing to Support a Beam

You can support a beam by pushing it up from below or by pulling it up from above. The support should be greatest at the point of greatest stress. In the next Activity, you will measure and compare the difference between the load that can be carried by a simple beam bridge and the load that can be carried by a bridge that is supported.

Support Your Beam

Problem

How much more load can be carried by a supported beam than an unsupported beam?

Materials

2 strips of thin cardboard (about 45 cm × 15 cm)
2 small boxes or other supports about 15 cm high
large plastic bottle with wide mouth
water
balance scale

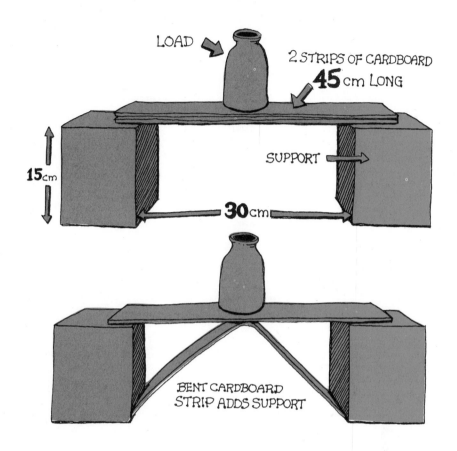

LOAD

2 STRIPS OF CARDBOARD **45** cm LONG

15cm

SUPPORT

30cm

BENT CARDBOARD STRIP ADDS SUPPORT

Procedure

Note: Do this Activity near a sink and be ready to clean up water if your design fails.

1. Place the two supports about 30 cm apart.
2. Make a simple beam bridge by placing the two strips (beams) of cardboard, one on top of the other, across the boxes. Place the plastic bottle on your bridge.
3. First make a prediction. How much load (mass) do you think your bridge can carry before it starts to bend? To find out, do a load test. Pour the water *very slowly*, just a little bit at a time, into the plastic container on your bridge. As soon as the bridge starts to bend, stop pouring.
4. Measure the mass of the water and the bottle.
5. Now, remove one of your cardboard beams. Use this strip to support the remaining cardboard beam, as shown in the illustration.

6. Predict how much mass your newly designed bridge will support before it bends. Now, test it!
7. Measure the mass of the water and the bottle, and record your findings.

Finding Out

1. How much additional load could your bridge support after you redesigned it?
2. Did your design add to the tensile strength or the compressive strength of the bridge? Explain.

Extension

3. How could you support a beam bridge by pulling it up from above?
   (a) Sketch a plan for such a bridge.
   (b) What additional materials would you need?
4. Develop a new design of your own that uses one strip of cardboard as a beam, and another strip as a support for it. Predict and then measure the strength of your new design.

Only One Support

A structure such as a diving platform looks like a simple beam. But there's a difference. The diving platform is supported only at one end. This type of structure is called a **cantilever**. When a load is placed on a cantilever, it does not bend in a U-shape like a beam. It bends the other way in a ∩-shape. Therefore, the top half is in tension while the bottom half is in compression—the opposite situation to that of a beam with a load on top. Because there is only one support, a cantilever must have high tensile strength. Few materials have this property. In fact, there were no suitable materials for building long cantilevers until researchers developed special alloy steels early in this century.

A cantilever structure often supports a balcony in a theatre or cinema. The cantilever avoids the need for columns, which might block the view of people below the balcony.

The first all-steel cantilever bridge, built in 1890, was the Forth Railway Bridge in Scotland. This diagram was drawn in 1890 to show how the cantilever bridge supports a load. The man in the middle represents a load and is seated in the centre of the span between two cantilever arms.

A cantilever railway bridge to span the St. Lawrence River was begun in 1904. It was designed to have a span of 550 m. The bridge collapsed as soon as it was completed, dropping 20 000 t of steel into the river and killing 75 people. This disaster was the result of errors in calculating loads and stresses and in assembling the parts.

The bridge was redesigned and rebuilt. The centre section was floated into place and attached to the ends of the cantilever spans that were built out from each bank of the river. As the centre section was being raised, part of the lifting assembly broke off. The centre section broke before falling into the river. This time the disaster was caused by one badly made piece of material.

Soon after, a new span was built. It was successfully lifted into place, and the bridge was opened in 1918. It has been in use ever since, and remains the longest cantilever bridge span in the world!

Cranes are a kind of cantilever. They are widely used in dockyards and on construction sites. Cranes use a cantilever structure to pick up and move heavy loads.

A cantilever bridge across the St. Lawrence River in central Canada.

One of the most common cantilevers is the wing of an aircraft.

Designing to Strengthen a Beam

The size of load that a beam can support without bending depends partly on the **orientation** of the beam. In other words, it depends on how the beam is placed. You can see this easily by looking at the beam of thin wood shown in the diagram. When the beam is placed on its narrow edge, it can support a much heavier load. That is because the load is being supported by a greater thickness of wood.

Try bending the flat side of your ruler slightly. Now try bending its narrow edge.

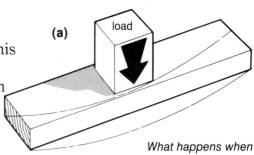

(a)

What happens when a load is placed on the wider side of the beam?

Less Is More!

Does a beam carry a load when there's nothing on it? That may seem like a trick question. In fact, the beam *is* carrying a load—its own mass.

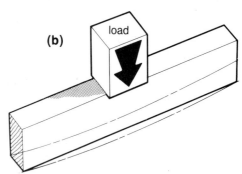

(b)

What happens when a load is placed on the narrow edge of the beam?

Activity 2-12

Is All This Beam Really Necessary?

Problem

Can you reduce the mass of a beam without weakening it? The diagrams show a beam, and the same beam after it is bent under a load.

- Measure and record the distances AB, CD, and EF on the beam when it is straight.
- Lay a piece of string along the lines AB, CD, and EF on the beam when it is bent under the load. Mark the ends of the lines on the string. Measure the string to find the lengths of the lines. Record your measurements.

Finding Out

1. When the beam is bent,
   (a) which line is under tension?
   (b) which line is under compression?
2. What is the effect on the length

(a) of tension?
(b) of compression?
3. Is there a line in the curved beam that is neither under compression nor under tension? How do you know?

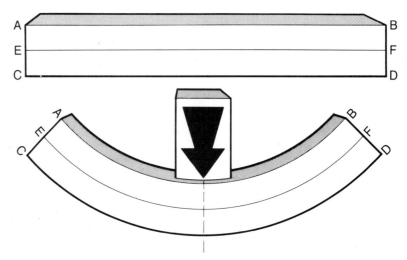

The Halfway Line

As you can see, there is a line along the centre of a beam that is neither squashed nor stretched when the beam is bent. That means this part of a beam receives no stress from a load. This line, which is exactly halfway between the top and bottom of a beam, is called the **neutral axis**. Along the neutral axis, the beam is doing very little to support a load. Therefore, the material in the centre of a beam can be reduced without weakening the beam.

This fact is used by builders and engineers in construction. Next time you pass a bridge or a construction site, take a close look at the beams. They may well be shaped like one of those shown here. I-beams, L-beams, and box-beams have less mass than solid beams. They are therefore less expensive. And with less mass of their own to support, they can support a greater load.

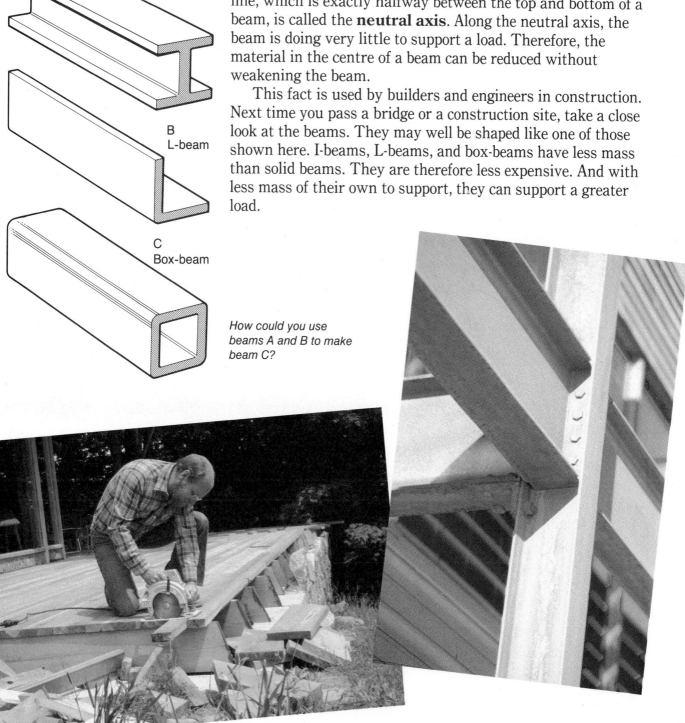

A
I-beam

B
L-beam

C
Box-beam

How could you use beams A and B to make beam C?

1. Copy each of the materials listed below into a table. Beside each, write down an important property of the material. In a third column, write a function that this material can be used for.

plastic	metal
rubber	plaster
wood	glass
paper	fabric
stone	

2. "Roberta and Willis built a swing from a piece of wood and a rope. They tied the rope to a sturdy branch. When Willis sat on the swing, the rope snapped. They got a stronger piece of rope. Roberta tried out the swing. This time, the wooden seat bent and then broke."

   (a) Identify the structure in this story, then tell the materials it was made of, and the loads that caused it to collapse.

   (b) Rewrite the story, using the words:
   load, stress, tension, compression

3. A farmer uses a plank bridge to cross a stream. The plank broke when a cow tried to walk across it. "It was never built to hold a cow," the farmer says. "I'll have to give it more support." "Or else," says a friend, "you can just make it stronger."

   (a) Describe how the farmer might support the bridge.

   (b) Describe how the friend might make it stronger.

4. A shelf is bending under a load.

   (a) Which side of the shelf is being compressed?

   (b) Which side is under tension?

   (c) How could the design have been improved to prevent the sagging? Draw a sketch if you wish.

5. You and a partner own a marina for boats. You have several rental boats. You want to buy a new rope for securing them to the dock. You've had bad experiences with ropes of various kinds, finding that many last only a short time.

   A rope supplier has shown you three different types of rope and would like you to decide among them. These are the facts the supplier has given you.

   (a) One type of rope is very strong, and is guaranteed not to break for 10 years. The cost will be $500 for the amount you need.

   (b) The second type of rope you have been shown is also guaranteed for 10 years so long as it is not kept in direct sunlight. The supplier has told you that its fibres sometimes break down in sunlight. The cost for the amount you need is $300.

   (c) The third type of rope this supplier has available is guaranteed for 5 years, and it may well last longer. The same amount of rope as above costs $100.

   Select one of the types of rope and write what you'd tell your partner were the reasons for your choice.

Building Up, Out, and Over

Jenny was asked to build a coat stand from a stick of wood about two metres long. She soon found out her biggest problem: How could she make the long piece of wood stand upright? She thought of several ways to do this. In the end, Jenny decided to add supporting pieces of wood around the base of the upright stick. The supports that Jenny used are called **braces**.

Jenny's solution for supporting a tall structure is one that is used commonly both in manufactured structures and in the natural world. A supporting brace, placed with one end on the ground and the other against a tall structure, forms the shape of a triangle with the ground and the structure. The triangle is a very strong and rigid shape. It cannot easily be bent. Test some triangles in the next Activity.

JENNY DECIDED ON A DESIGN. THIS IS WHAT SHE BUILT.

*The supports against the tree and the church are special kinds of braces called **buttresses**.*

Why Be a Square When You Can Be a Triangle?

Problem

How do triangles strengthen a frame?

Materials

strips of stiff cardboard
paper fasteners or pins

Procedure

1. Join four strips of stiff cardboard with pins to make a square frame. Push the frame at point A.

2. Add one long strip of cardboard to your frame in a way that will make it more rigid. Test the frame by pushing at point A.

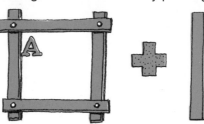 

3. Make another square frame. This time make it rigid by using four short strips of cardboard.

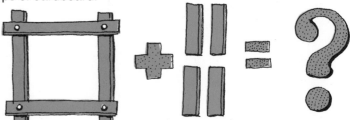

4. Make a rectangular frame. Make it rigid by adding two strips of cardboard.

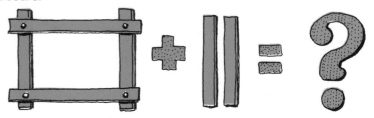

Joining Triangles Together

You have seen that you can make a square or rectangular frame much stronger by making it into two or three triangles. You can add even more triangles together side by side to make a framework that is lightweight but very strong. This type of framework is used in building tall structures, such as electricity pylons, or the Eiffel Tower in Paris. It is also widely used in many other structures. On your way to or from school, keep a lookout for triangles. You may be surprised at how often you see this shape once you start looking for it.

The Eiffel Tower, Paris, France.

The pyramids of Egypt.

The canopy of Sun Life Place, Edmonton.

Electricity pylons.

So far, you have looked at triangles joined together side by side. Triangles can also be stacked together, one behind the other, to make the shape shown in part B of the diagram. This shape can work like the beams you studied in Topic Four. Now imagine a series of these long shapes side by side. You can see this in sheets of **corrugated** metal that are used on some shed roofs, or in the corrugated cardboard used to make strong grocery boxes. When a sheet of metal or cardboard is shaped into a series of triangular folds like this, it becomes more rigid and stronger than a flat sheet. You can prove this for yourself in the next Activity.

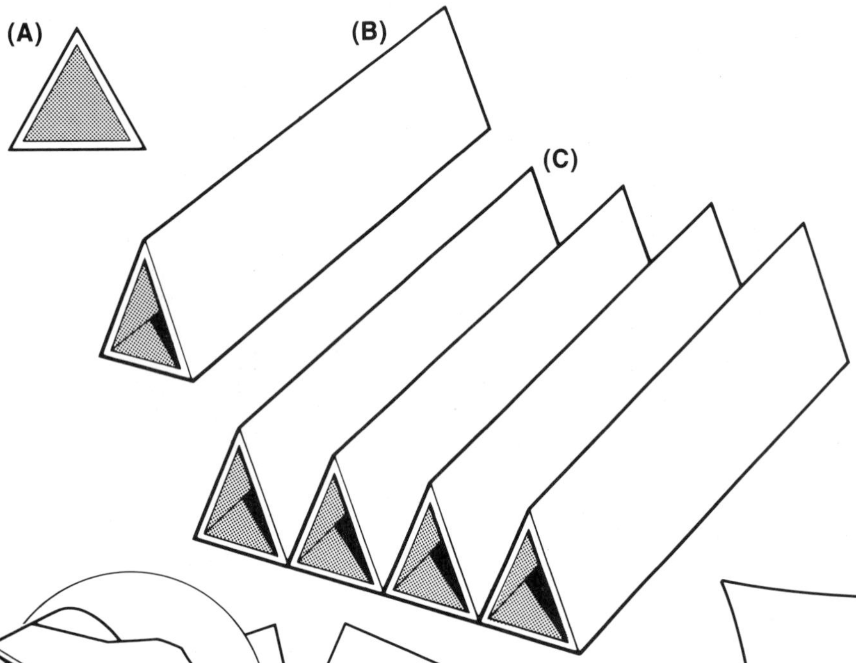

(A) (B) (C)

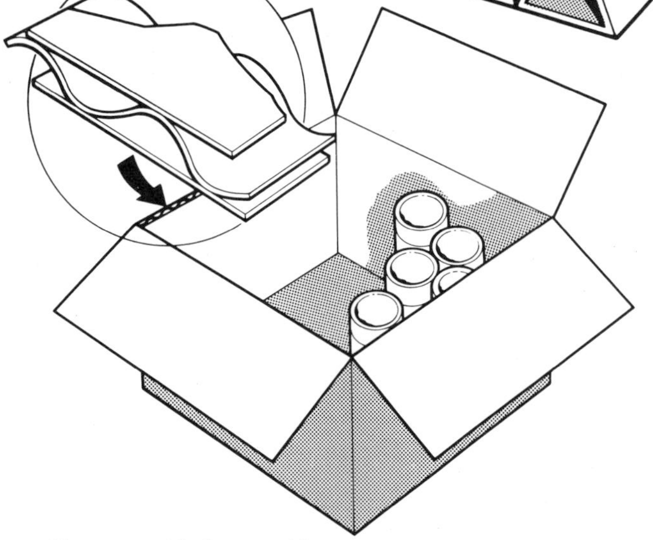

Boxes used to transport heavy cans of food are reinforced with corrugated cardboard.

Making Rigid Sheets

Problem

How can you make a flat sheet more rigid?

Materials

2 sheets of paper or thin cardboard (about 28 cm × 22 cm)
small (250-mL) plastic container
marbles or other small, heavy objects
2 small boxes

Procedure

1. Hold a sheet of paper by the bottom edge and try to make it stand upright. What happens to the sheet of paper?
2. Fold the sheet lengthways in half. Now try to stand it upright. What is the difference between this design and the one above?

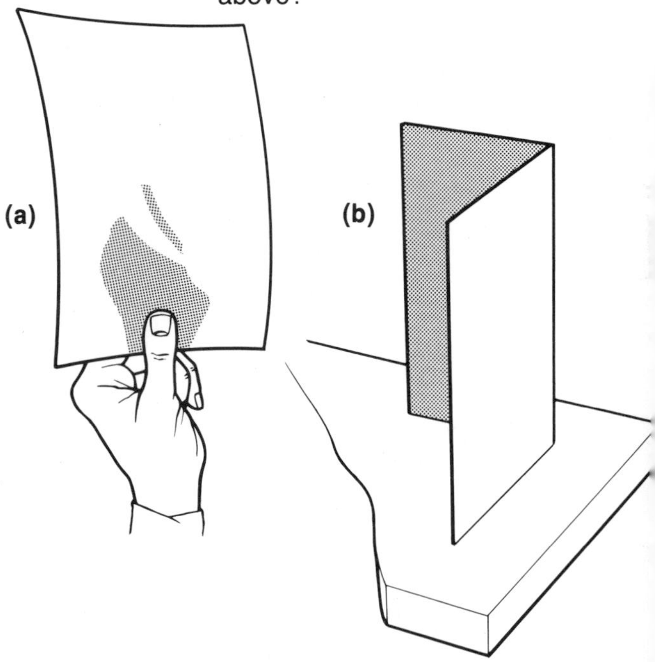

(a) (b)

3. (a) Place two small boxes about 15 cm apart. Place a sheet of paper so that it makes a bridge between the two boxes. Will this bridge support an empty plastic container? Try putting one or two marbles in the container. What happens?
(b) Make five folds in the sheet of paper to make it more rigid. Will this bridge hold the container and marbles?
(c) Test the strength of your bridge by adding more marbles to the container until the bridge starts to bend.

Finding Out

Can you make your bridge stronger by adding more folds? Try some different designs and test them by using the marbles and container as a load.

(c)

(d)

"Waveboard" is a new kind of wooden panel developed by the Alberta Research Council. It looks like corrugated cardboard. It is made by mixing thin pieces of wood with sticky substances called resins. This mixture is heated in a special mould to form the wavy panel. Waveboard is much stiffer and can support 10 times as much load as flat wooden panels like plywood. It also has less mass.

"Stressed skin panels" are made from waveboard placed between two wood "skins." Both waveboard and stressed skin panels can be used in construction instead of metal or flat wooden panels.

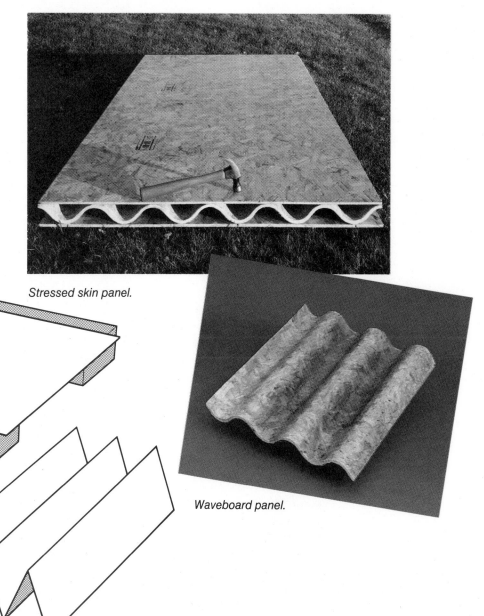

Stressed skin panel.

Waveboard panel.

Columns

You learned at the beginning of this Topic that an upright stick can be supported by braces at its base. If an upright structure already has a broad base, however, it can stand up by itself without extra support. This kind of structure is called a **column**. Columns were often used by the ancient Greeks as a simple way to build upwards. A building could be made of several columns in a row, with beams running across the top of them. These Greek buildings look very beautiful. But building with columns poses several problems.

The Parthenon, Athens, Greece.

The Temple at Philae, Egypt.

Stonehenge, Salisbury Plain, England.

Columns are also used in new buildings: college at Grande Prairie, Alberta.

1. A crossbeam supported on two columns is similar to the bridge you studied in Activity 2-11. As you discovered, a structure like this will bend or break if a heavy load is placed on it. Therefore, you cannot add a large mass above the centre of the crossbeam where it is not supported by the columns. This makes it difficult to add a large second level to a building constructed from columns.

So Far and No Farther

Place a strip of cardboard across two supports a short distance apart. Put a plastic container on the cardboard and add marbles until the beam can just support the load without bending. Now move the supports farther away from each other. What happens?

2. A long crossbeam cannot support as large a mass as a short crossbeam. Therefore, columns cannot be built too far apart, or the crossbeam may break under the load. You can test the relationship between length and ability to support a load in the next Activity.

3. Two columns standing on the ground, with a beam across the top, make the shape of a rectangle. As you saw in Activity 2-13, a rectangle that is not braced can easily be pushed to one side. Therefore, a structure built in this way is not very rigid.

Arches

Over the years, people have thought of a number of solutions to the problems that columns pose. One is to use an **arch** to support the crossbeam. The arch is one of the simplest and strongest structures for supporting a load. The Romans used arches in their buildings over 2000 years ago. Arches built by the Romans are still standing and in good condition today.

The design of the arch allowed early builders to span wider openings. They built up an arch with small blocks of brick or stone. They had no need for mortar. The building blocks in the arch were wedge-shaped, and pressed closely against one another. This shape conducts the stress of the load from the top of the arch down onto the ground. For this to work, however, the base of the arch must be firmly anchored.

Do you remember the arch you constructed in Activity 2-11 to support a beam bridge? By using arches, builders were able to construct much larger and stronger structures than ever before.

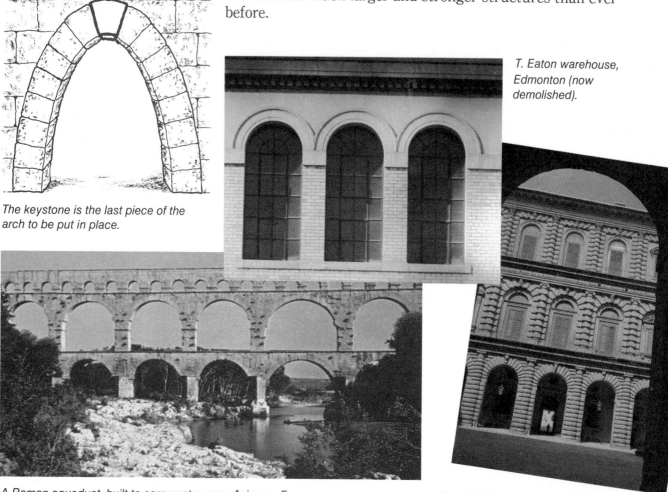

The keystone is the last piece of the arch to be put in place.

T. Eaton warehouse, Edmonton (now demolished).

A Roman aqueduct, built to carry water, near Avignon, France.

The Pitti Palace, Florence, Italy.

A Case Study: Just an Ordinary Bridge

Building any large structure involves many decisions. Planners must consider factors such as:

- the design;
- the materials to be used;
- the cost;
- public safety;
- the appearance of the structure;
- the length of time it will take to build.

To understand how some of these decisions are made, read this case study. It describes the steps the planning team had to take before building the new Devon Bridge.

For many years, a two-lane road bridge carried traffic over the river to the town of Devon, Alberta. During the 1970s, more businesses and people moved into the area, and the amount of traffic using the bridge increased. This led to traffic jams and many traffic accidents, especially in winter on the steep, icy slopes leading down to the bridge. It has been decided to build a new, wider four-lane bridge, one that will carry traffic from the tops of the banks on either side of the river.

There are many different types of bridge design, but the planners have decided that the new bridge should consist of **girders** (long beams) resting on **piers** (support columns). At this point, the planning team must make two further decisions: what materials to choose, and what design to use.

The old, two-lane Devon Bridge on the left had to be replaced. The steep slopes down to the river on either side were hazards in winter. The new Devon Bridge is under construction on the right.

Choosing the Materials

As studies show that steel and concrete girders are equally strong, how will the team choose between these two materials? Some of the points to consider are:

- Which material takes longer to assemble and join together?
- Does one material have to be transported a longer distance to the bridge site than the other?
- Are the construction workers in the area more experienced using one material than another?
- Which material is more expensive?
- How much of each material would be needed for the bridge?

When the costs are added up, the team finds that to use concrete will cost over a million dollars more than to use steel. For this reason, the planning team decides to use steel for the new Devon Bridge.

Choosing the Design

A bridge can be built with many combinations of girder lengths and numbers of piers. If the girders are long, the bridge will need fewer piers. If the girders are short, the bridge will need more piers. How can the team decide on the best combination of girders and piers? Here are a few things to consider.

Advantages of long girders
- Long girders look better than short girders.
- Long girders mean fewer piers.

Disadvantages of long girders
- Long girders have more mass, which increases the load on the piers.
- Long girders are more expensive to manufacture.
- Long girders are more difficult and more expensive to transport to the construction site.

Number of piers
- Piers are expensive to build.
- Piers take a long time to build.
- Cold weather during winter can hold up pier construction.
- Building piers can be dangerous, so the more piers there are the greater the risk to workers.

There is no such thing as a perfect design. The planning team must find the design that has the best combination of cost, safety, speed of construction, and attractive appearance.

The example of the Devon Bridge shows some of the typical questions that must be asked and answered when communities want to build large structures.

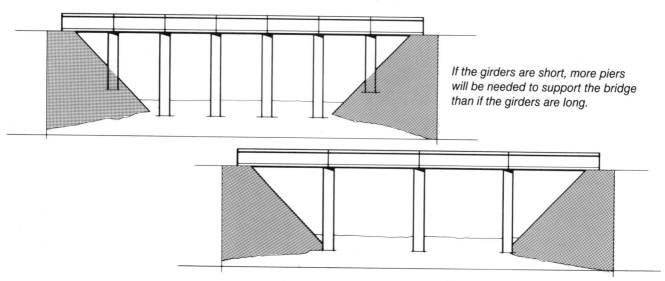

If the girders are short, more piers will be needed to support the bridge than if the girders are long.

Most of the decisions must be made well before the first steps in actual construction begin. Even the construction itself must be planned well in advance. For example, you would not want to have the girders manufactured and delivered several weeks before the piers were completed. Based on past experience, people have developed design codes and safety standards. These guidelines help planners to make their choices.

Finding Out

1. What issues led people to decide that a new bridge was needed across the river?
2. Once the decision was made to build a new bridge, give at least three problems the planners and engineers had to debate and solve before starting to build the bridge.
3. Give the ideas or plan they had for solving one of these problems.
4. What were some of the factors they had to evaluate in their building design?
5. You have been testing the designs and structures you have made in this Unit. Why do you think the planners of the new Devon Bridge did not test their design?

Finding Out More

6. The design and materials used to build the new Devon Bridge were good choices for that particular situation. Describe a situation in which you feel a different design or different material would have been selected. Use your imagination!

The new Devon Bridge was completed in 1988.

Making Connections

The strength of a structure depends on its design, shape, and the materials used. The strength of a structure also depends on the way its parts are held together. The point at which parts are connected is called a **joint**.

A poorly-made joint is often the first part of a structure to break under stress. Therefore it's important to find the best possible way of joining one part to another. In most cases, the joint should be as strong as any other part of the structure.

Rigid Joints

Some joints connecting parts of a structure must not move. A joint that does not allow movement is a **rigid joint**. How the parts are connected to form a rigid joint depends on

- the materials of which the parts are made; and
- the way in which the structure is to be used.

If you had to join together two bricks, how would you do it? Mortar is the obvious choice. But sometimes the best method of forming a joint isn't so clear.

Maybe I should have used nails and wood screws too!

PAPER CLIPS

GLUE

ELASTIC BANDS

This is a model of a two-seater wooden aircraft. The original was built over 70 years ago.

122 Structures and Design

For example, parts made of wood can be joined together using nails, screws, or glue. The best method depends on how the structure is to be used.

Designers of wooden aircraft found this out during the early 1900s. To join the parts of the aircraft together, they used nails and screws—methods similar to those used in making wooden furniture. Unfortunately, the stresses placed on an aircraft in flight are not the same as those placed on an armchair in your living room! Vibrations, strong winds, and other stresses tore apart aircraft parts held together by nails or screws. The designers quickly developed a system of joining the parts securely with glue.

Activity 2-17

Making Strong Joints

Problem

How can you measure and compare the strength of different rigid joints?

Procedure

1. This is an experiment that you can design for yourself. Use the materials shown here to design four different ways to join together pairs of sticks. You can use more than one material in a joint if you wish.

2. When you have made your four joints, predict which joint is likely to be strongest and which is likely to be weakest. Develop a method to test the strength of your joints and see if your predictions were correct.

3. Check with your classmates to see how many different kinds of joints were made. Which were the most successful designs?

Join your sticks together like this.

Mobile Joints

What would happen if your thigh bone were joined rigidly to your shin bone? It would be like having your leg in a cast—you would find it difficult to move around! Your knee is an example of a **mobile joint**. A mobile joint allows parts of the structure to move. But the joint must still be strong and able to withstand stress.

One function of a mobile joint is to allow part of a structure (not the whole structure) to move easily in response to a light load. A door, for example, can be thought of as part of a wall joined to the rest of the wall by a mobile joint—the hinge. The door allows people to pass through a wall more easily than if they had to knock a hole in it and reassemble it behind them each time! An umbrella has mobile joints that allow you to change its shape. The caterpillar tread on a construction vehicle is made up of many hinges attached together.

There are many mobile joints in the natural world. For example, shellfish such as clams have hinges to open and close their shells in order to feed and protect themselves. A plant, the Venus flytrap, uses a hinge joint to open and close its leaves to trap insects. Without other kinds of mobile joints similar to the joint in your knee, you and many other animals would not be able to move around.

Both the bicycle and the human body include parts joined by rigid joints and parts joined by mobile joints. Identify all the joints you can find in each structure. Then list the joints in one structure that are similar to joints in the other.

L. SENIN

Checkpoint

1. Which of the following shapes is more rigid? Explain.
   (a) □
   (b) △
2. How could you make the less rigid of the two shapes above more rigid?
3. (a) What does *corrugated* mean?
   (b) Write a poem, short story, or cartoon that explains why you think someone invented corrugated cardboard. Feel free to make the problem that led to the invention funny.
4. An architect wants to construct a building using several columns with crossbeams running across their tops. He wants the columns to be 50 m apart. He also wants to build giant stone statues halfway along each beam.
   (a) What problems might result from this design?
   (b) Describe two ways of solving these problems.
5. Explain what is wrong with these statements:
   (a) "As long as I can join these two pieces firmly together, it doesn't matter what I am going to use them for."
   (b) "This new super-glue is the best way to join any two things together."
   (c) "All joints must be firm, rigid, and strong."
6. What examples can you find of various kinds of joints? This is a research project. First, look around at home and school for structures made up of parts that are joined together. (The structures can be natural or manufactured.)

(a) Make a table to record your results.
(b) Name the two parts that are joined together, and the materials they are made of.
(c) Record if the parts are joined by a rigid or a mobile joint.
(d) Describe the method used to make the joint.

Make sure you find at least one example of each of the following:
- a rigid joint between two similar materials
- a rigid joint between two unlike materials
- a mobile joint between two similar materials
- a mobile joint between two unlike materials

7. The diagram shows a shear pin in the propeller of an outboard motor. The shear pin is a joint that has been designed to be weak.
   (a) What will happen if the propeller strikes a submerged rock?
   (b) What advantage is there in having a weak joint in some structures?

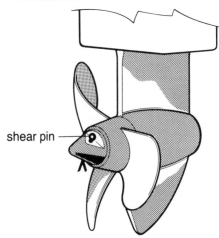

shear pin

Science and Technology in Society

A Helping Hand—and Elbow—for Kids

When someone is born without a limb or loses one in an accident, he or she may need an artificial limb. Designing an artificial body part presents a real challenge. The designer wants it to work as much as possible like the part it replaces. Suppose you were designing an artificial hand and arm for a child. Here are some of the things you would have to consider.

1. *How should the arm be moved?* A natural arm and hand have many different joints, which are moved by muscles. What could be used instead?
2. *How heavy can it be?* It must be made of lightweight materials, or it will be too awkward for the child to use.
3. *How strong should the materials be?* The child needs the arm every day, so it must work without breaking down.
4. *How should it look?* It should match a natural arm in shape and colour.

Here is one solution to these problems. This child-sized device includes an artificial elbow, lower arm, and hand. It was designed in Canada, at the Hugh MacMillan Medical Centre.

This artificial hand operates *myoelectrically*, using muscles and electricity. (*Myo* comes from the Greek word for "muscle.") Just before a person flexes a muscle, it gives off a tiny burst of electricity. The myoelectric hand can use this electricity. The child learns to flex a muscle that sends an electric signal to the hand. Sensitive switches in the hand pick up the tiny signal and transmit it to a small, battery-powered motor. This motor then makes the hand move.

The hand has a thumb and two fingers which can meet in a strong pinching movement. A soft vinyl glove, matched to the child's skin colour, covers the hand. The glove has a full set of fingers with outlines of nails and knuckles. Unless you look closely, it's hard to tell the glove from a natural hand. The hand can open and close about 500 000 times before it needs repairs.

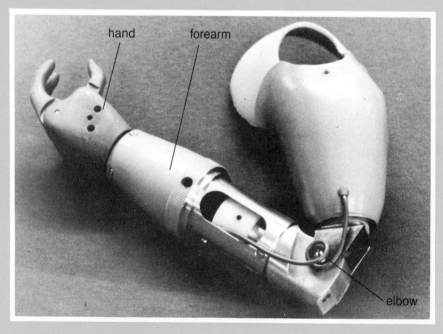

This artificial arm is for an eight-year-old child.

For children who need an artificial arm as well as a hand, the Hugh MacMillan team has designed a lightweight elbow. The elbow has a hinge joint so that the arm can bend. Just above the hinge is a small turntable, so that the arm can also move from side to side.

Some children can use their shoulder muscles to operate the elbow, like the hand, myoelectrically. Other children may need to make the elbow move by using push-button switches on the outside of the elbow. Even if these children have to use switches for the elbow, they can still use their muscles to operate the hand attached to the elbow.

Think About It

Why does the myoelectric hand have only two working fingers and a thumb? You have an *opposable thumb*. This means that your thumb faces your fingers. Almost no other animal has a limb like this.

1. Tape your thumbs to the palms of your hand so that you can't use them.
   Try to:
   Write your name.
   Do up some buttons.
   Tie your shoelaces.
   Pick up a fork.
   Throw a ball.
2. Now tape your little finger and your ring finger against your palm. Try the activities above, using just two fingers and a thumb on each hand.
3. What happened when you tried to do the activities without your thumbs?
4. What happened when you tried to do the activities with your fingers taped?
5. Why do you think the myoelectric hand is made with two fingers and a thumb?

Laurin Nightingale is only two-and-a-half years old. She is using the world's smallest myoelectric arm to assemble a toy.

The myoelectric hand on the left was designed for a child between the ages of two and six. The hand on the right is for a fifteen-month-old baby.

Putting It All Together

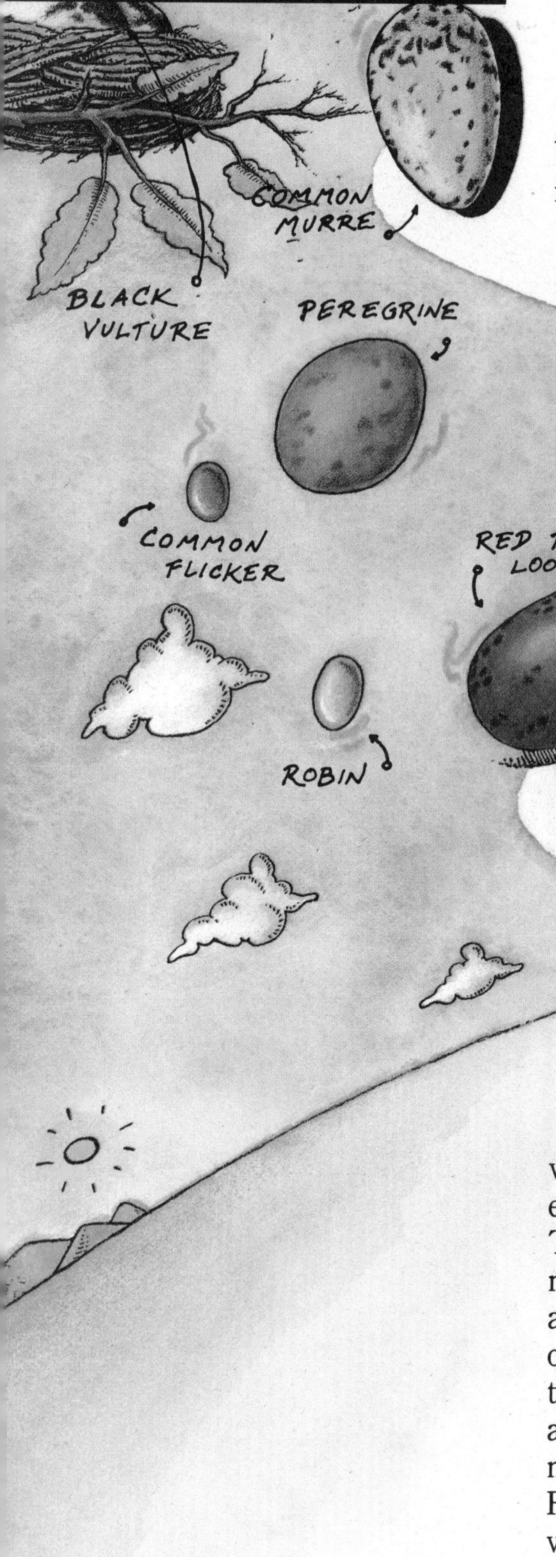

What makes a well-designed structure? At the very least, it must

- perform the function required of it, such as providing shelter or support;
- be made of the most suitable material in the most economical amount;
- be as strong as necessary to resist the stresses that will be put on it;
- meet any required safety standards.

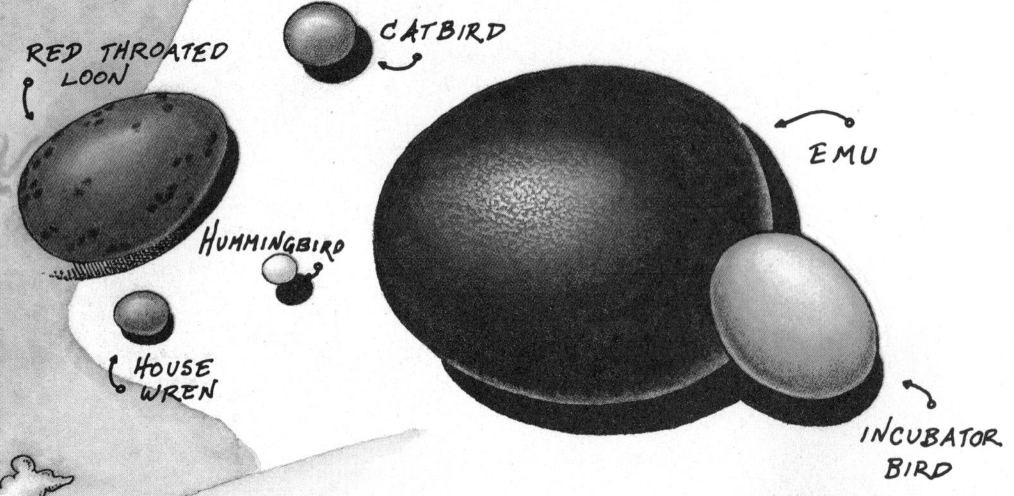

COMMON MURRE

BLACK VULTURE

PEREGRINE

COMMON FLICKER

RED THROATED LOON

CATBIRD

EMU

HUMMINGBIRD

ROBIN

HOUSE WREN

INCUBATOR BIRD

CEDAR WAXING

An egg is a good example of a well-designed structure. It completely encloses and protects the growing chick. The shell is made of a hard, resistant material that keeps water out and lets air through. The oval shape allows the shell to enclose a fairly large amount of space with a minimum of material. Finally, an egg is very strong.

Treading on Eggshells

Problem

How much load can an eggshell support?

Materials

four eggs
small pair of scissors
masking tape
fine sandpaper (optional)
several large books about the
    same size (encyclopedia
    volumes are good)
bathroom scale

Procedure

1. Over a sink, gently tap one end of each egg to break the shell. Carefully cut away a little of the broken shell to make a small opening. Pour the contents of the egg out through the hole.

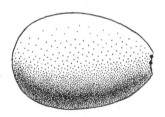

2. Stick a length of masking tape around the centre of each shell. This will prevent the shell from cracking when you cut it.

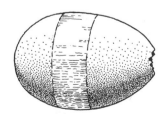

3. Carefully cut around each eggshell, through the masking tape. You should end up with four half-shells of equal size (halves without the holes in them), with level edges. If you want, you can gently sandpaper the cut edges to make them smooth and flat.

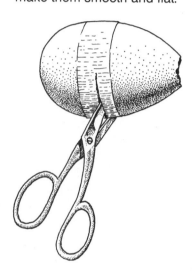

4. Stand the eggshells on a smooth, clean floor or tabletop to make a rectangle slightly smaller than the size of your books.

5. Carefully place a book flat across the eggshells. Add another. Be sure the shells do not move when you add each book. The load must always press directly down. If you push or twist on the shells, they will break.

Finding Out

1. What total mass did the shells support?
2. What do you think happens to the compressive stress at the top of the shell?

"Egg-zactly" Like a Dome!

An egg is shaped like two **domes** placed together. A dome can be thought of as a three-dimensional version of an arch.

Like an arch, the dome design is very strong. It is used in large, spacious buildings such as cathedrals and modern sports stadiums. The dome shape distributes the load of the roof through the dome's curved walls to the broad base. This shape has the added advantage of not needing columns to support the roof. Columns would block the spectators' view!

One problem with large domes is that the mass of the material pushes down and out near the base. This causes the dome to spread out. For this reason, many older domed buildings have very thick walls for support, or they use buttresses to hold in the walls against this outward-pushing stress. (Look back at the cathedral on page 111. Notice that the buttress on this cathedral is itself in the form of an arch.)

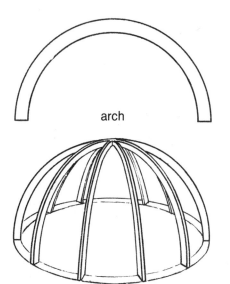

arch

dome

A dome is a three-dimensional structure. It is like a series of two-dimensional arches all joined at the top.

Zulu huts in South Africa.

The strength of the dome design has made it widely used at many times.

Mosque of Ibu Tulum in Cairo, Egypt, built over a thousand years ago.

B.C. Place, Vancouver.

Did You Know?

An ingenious variation on the dome design is the **geodesic dome**. This structure was developed in the 1940s by the American architect Buckminster Fuller. A geodesic dome is built up from a framework of triangles, usually made of steel or aluminum tubing. Triangles, as you have seen, distribute stresses evenly. The stresses in a geodesic dome all remain within the walls of the structure itself. Thus, a geodesic dome can be set directly on the ground as a complete, self-contained unit that does not need any extra support.

St. Peter's Basilica, Vatican City, built in the 1500s.

A geodesic dome uses the structural advantages of the dome shape and the triangle shape.

Reinforced concrete dome, Calgary Planetarium.

Super Straw Structures

Strawville needs a centrepiece for the town. The town council is holding a competition for the design and construction of this structure and has invited you to take part. The centrepiece can be anything you like. It might be a tower or a monument. (If you're really ambitious, it could even be a community centre!)

To show the people of the town what your building will be like, you must construct a model. The Strawville planning committee wants the model to have these features:

- It must be built only of plastic drinking straws, joined in any way you choose.
- It must be at least two straw lengths in height and at least two straw lengths along one side.
- It can be in any shape you choose, but it must be tested for its ability to support a load without bending. The load must be at least as great as the mass of a tennis ball.

Awards will be given for
- the tallest model
- the strongest model
- the most imaginative model
- the model that uses the fewest straws

Procedure

1. Work in groups. Use what you know about structures to build the best model you can. Here are some terms to remind you of ideas you've learned: compressive strength, tensile strength, beams, columns, arches, triangles, braces (buttresses), domes, loads.

2. Start by planning the shape of your structure. Design the frames you will use for your structure. What sizes will they be? How will you make them strong and rigid?

3. Remember to test and evaluate your structure as you go along. Consider alternatives, and make any changes you need to improve your structure.

4. When your class has finished, choose a member of your group to present your model to the planning committee. The planning committee will vote on which model should get each award. Then decide on a model in each category you think Strawville should choose as a runner-up. What factors did you consider in selecting the runners-up?

Push one end of a clip through the cup. Bend the clip into a hook.

Squeeze one end of a straw and push it inside another straw to make a rigid joint.

Slide another clip onto the first.

Push onto the clip to make a joint at the ends of the straws.

Push the side of the clip that has two ends into the end of the straw.

Push the straw through the clip to make a joint along the length of the straw.

Hang the cup from different parts of your structure. Add washers, coins, or nails until the straw bends. If necessary, strengthen the part and try again.

Here are some suggestions on how to join your straws together.

Building in Space

Designing for Space

Even as short a time ago as the early 1950s, there were no human-made structures outside of the Earth's atmosphere (the envelope of gases surrounding the Earth). That all changed in 1957 when the Soviet Union launched a rocket that carried the world's first orbiting space satellite, *Sputnik I*. Within 10 years, more than 1200 satellites and other vehicles had been launched into space!

The environment of space is very different from the environment we are used to on Earth. Structures that are sent out into space are designed to meet these different conditions. For example:

What similar needs do deep-sea divers and astronauts have?

- There is no weather in space—no wind, rain, or snow. Once they are in space, structures need not resist wind stress, snow loading, or rust.
- When spacecraft speed through the Earth's atmosphere after being launched, or when they return to Earth, they become very hot. They must therefore be made of material that can be heated without melting. The material must also *insulate* the craft. That is, it must not allow the heat to pass through the walls of the spacecraft to the inside, where it would harm the crew or delicate instruments.
- Because it is difficult—and expensive—to launch structures with a large mass and volume, designers of spacecraft must try to keep their structures small and lightweight.

In some ways, the environment of space is similar to being underwater. Think about how these two environments are similar.

Space Stations

One important goal of space programs—and the dream of many science fiction writers—is to build large stations in space. People would live and work in these stations for long periods of time.

All the parts for a space station must first be built on Earth and then sent into space in pieces. Then the space station can be assembled beyond the pull of the Earth's gravity. (You will learn more about gravity in the next Unit.)

Working in space isn't easy. Therefore, to put together a station in space, pieces of a space station must be designed to be fastened tightly together using a minimum of materials, tools, and effort. The Soviet Union's *Soyuz* was the first space program to weld metals together in space.

A space station that will house a crew for long periods of time must be designed to be completely self-contained. It must have within it all the oxygen, fuel, and food needed. It must also have a way to dispose safely of wastes. To reduce the bulk and size of equipment that has to be launched from Earth, space stations are designed to generate their own energy (from sunlight) and grow their own food (using plants).

A plan for a space colony between Earth and the Moon in the 21st century. The two cylinders are 32 km long and 6400 m in diameter. The cylinders are living areas. This space colony structure is designed to hold from 200 000 to several million people, depending on how the inside is planned. Moon or asteroid materials would be used for construction. Large moveable mirrors would direct sunlight into the interiors, regulate the seasons, and control the day-night cycle.

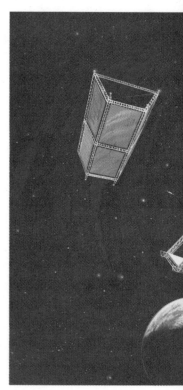

Working in Space

The trips taken in space so far have been relatively short. People living in space for any length of time will have some problems to solve. For example, people who work in space, like people at home on Earth, are likely to get sick. By the late 1990s, there is a plan to include an emergency medical unit in an orbiting space station. However, designing the unit is difficult, because there will be so little room for it. All the facilities of a physician's office must be designed to fit into an area that's only as large as two telephone booths!

Activity 2-20

Designing for Space

In small groups, plan the design of a space station. Consider first where your station will be (on a planet or orbiting in space). How big will it be? How many people will work there? What facilities will you need inside it?

To plan your space station properly, you'll need to do some research on the conditions found in the part of space where your station will be. Using this information, decide what properties the materials used in your space station must have. What is the best shape for your station? Will there be any need for mobile joints (so that, for example, there could be an arm on the outside of the station)?

Consider the inside of the station. Will there be open spaces or many different rooms? Remember that the more material you use, the more costly and time-consuming it will be to launch and assemble the station. On the other hand, you must consider the functions of the station, and design it for those functions.

When your group has collected enough information and worked out these details, prepare a presentation for the rest of the class. You may choose to make a model of your space station. Or you may want to write a description and/or sketch your designs.

An artist's concept of a solar satellite power system. This is one of a number of drawings based on recent studies of possible future space activity.

Working by Making Things

Do you like taking things apart to see how they work—and putting them together again? If so, you are probably someone who enjoys making things.

An Eye for Design

Eric Kiisel was always interested in mechanical things and how they worked. He especially liked drafting and machine shop in high school. Then he went to college to study mechanical engineering technology. The work was interesting, but Eric realized he would rather design new machines than operate existing ones. He took further courses in mechanical engineering, physics, and mathematics, and is now a developmental engineer.

One machine Eric Kiisel has designed is a new kind of cutting tool. This machine uses high pressure jets of water instead of a solid blade to slice through steel or to cut elaborate shapes out of delicate plastic. Eric is excited by the possibilities the water jet cutter offers. He plans to use a robot, controlled by a computer, to operate the cutter. The cutter will then be able to cut exact shapes over and over again. This technique will be especially useful in making high quality aircraft parts. Eric is looking forward to showing others how to use the improvements he has designed and tested.

A Structure for the Cold

Domes are structures of great strength. At the University of Calgary, Dr. Peter Glockner and his students and co-workers have designed, built, and tested a reinforced ice-dome that could be used in the extreme cold of the Arctic. They inflated a shell of spun fiberglass to the shape of a dome. Then they sprayed it with water. At temperatures of ⁻16°C or even lower, the water quickly freezes, covering the dome with a fine layer of ice. Repeated sprayings build up the thickness of ice on the dome, and make the structure very strong.

Northern communities will be able to use these domes for temporary or semi-permanent coverings for storage or work areas. Peter Glockner hopes that these domes may even provide northern communities with domed roofs for skating rinks and other sports and recreation buildings.

(a) *A test model of the ice dome.*

(b) *Even a small ice dome can support quite a load.*

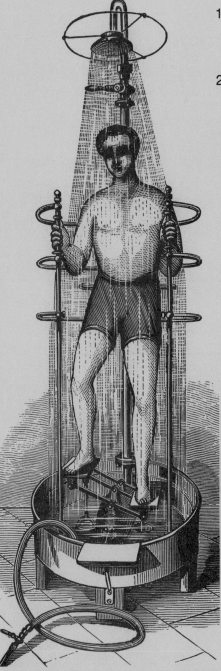

1. What shape of building would you choose when designing a large sports stadium? Explain your choice.
2. (a) A well-designed structure must meet four requirements. List these requirements.
(b) Look at the two inventions in the illustrations. Explain why each of them failed.
(c) List any features of a well-designed structure that each has.
(d) Choose one of the inventions and describe how you would change it to improve its design.
3. Research to find out why astronauts must wear special suits when working in space.

At last—a machine to do two jobs at once! Mereney's washing machine, invented in 1882, washed clothes and boiled grain and vegetables at the same time. Unfortunately, its tendency to explode, tossing clothes, grain, and vegetables on its user, reduced its popularity.

A shower that allowed the user to pump water by simple foot movement sounded like a great idea—especially for people living in an area where it was difficult to pump water in another way. Alas, most people found it too difficult to generate enough "pedal power" to get more than a trickle of water. Another inspired invention bites the dust.

Unit Focus

- A structure is an object made up of one or more parts. A structure can occur in the natural world or be made by people.
- The design of a structure involves the shape of the parts and the way the parts are put together.
- There are similarities between some manufactured structures and structures in the natural world.
- One of the most important functions of a structure is to resist stresses that might make it collapse or break apart.
- A particular task can usually be carried out by a number of structures with different designs.
- Materials have different properties. The materials of which a structure is made help the structure carry out its functions.
- The ability of a structure to resist being squashed is its compressive strength; the ability to resist being pulled apart is its tensile strength.
- Structures can be designed for strength.
- The parts of a structure are joined together by joints that are rigid or mobile, depending on their function.

Backtrack

1. Solve the crossword puzzle by using some of the words found in this unit.

2. Explain why you might use each of the following in a structure: beam, cantilever, arch, dome, column, brace, hinge.

Across Clues

2. A joint that is not mobile.
5. A beam supported at one end only.
6. A beam under a very heavy load will do this.
8. This stress acts to squash a structure.
10. To fasten two parts of a structure together.
11. Half an eggshell is this shape.
12. This creates stress on a structure.
13. A simple but strong structure used by the ancient Romans.

Down Clues

1. A type of support.
3. An igloo is a dome made from this.
4. This stress acts to pull a structure apart.
6. Part of a structure used horizontally to support a load.
7. A very strong and stable shape.
8. A vertical part of a structure used to support a cross beam.
9. A type of dome invented in this century.

3. Why is an arch often a better shape for a bridge than a simple beam?
4. For each of the following properties, name one material that has that property: hard, soft, flexible, rigid, waterproof, lightweight.

Synthesizer

5. You are designing a table and chair for a young child's room. Explain what you must consider in your design, including the size and shape of the furniture, its functions, and the materials you will use for it.
6. Explain why designers use plans and models before building large structures.
7. (a) What are the advantages of a structure that is made up of many parts that are the same shape and size?
   (b) Why are many structures made up of parts that are all different from one another?
8. Sketch three different designs for a bridge.
9. You are sent to study old buildings in Europe. You observe that the walls near the base of some old churches are cracked and slightly bulging. What do you think has caused these problems?
10. Describe briefly how a knowledge of loads, stresses, materials, and joints would allow you to build a strong, attractive, and long-lasting structure to store books.

11. This scene shows a home about 5000 years ago. The technology these people invented served them well. With their stone tools, they were able to carve out a dugout canoe. Out of a single piece of wood or a single large rock, they were able to carve out not only devices for travelling, but also stools for sitting, and bowls for mixing.
    (a) Looking carefully at the scene, think of at least three materials that have been discovered in the last 5000 years to replace these materials.
    (b) Name a structure that has been invented or improved by using each of these new materials.
    (c) Tell why you think the new device for serving the same function is an improvement.
    (d) Describe some new materials that you, as an inhabitant of the Earth in the year 2050, will introduce to replace materials we use today.
    (e) Write an advertisement for one of your new materials, to be placed in the *Galactic Gazette* to introduce your new product to the people on Earth.

Forces and Motion

As you lie in your bed at night, just about to fall asleep, everything seems motionless. But it is not. Your lungs move in and out continuously as you breathe. The blood in your body is always in motion as it is pumped rhythmically by your heart through 96 000 km of blood vessels.

The Earth, on which your bed stands, is spinning like a top as it travels around the Sun at a speed of about 100 000 km/h. As well, our entire solar system is revolving at great speed around the centre of a huge cluster of bodies in space, called the Milky Way.

Motion all over the universe depends on forces. Without forces, there would be no motion. Understanding how forces act helps people to build ships that float and planes and space vehicles that fly. Forces also prevent motion. They act together to keep our buildings and bridges securely in place.

Explore this Unit to see how forces act on all things, both living and non-living, and how they affect you every day of your life.

Experiencing Forces

Objects move because something made them move. That something is known as a **force**. Forces start objects moving by giving them a push or a pull. But forces do not only start objects moving. By pushing and pulling, forces also stop, speed up, slow down, or change the direction of moving objects. You are already aware of many forces, perhaps more than you realize. The next Activity will let you see some forces in action.

Finding Forces

Look at the illustration on these pages. See if you can find evidence of:

(a) a pushing force: What is doing the pushing?

(b) a pulling force: What is doing the pulling?

(c) the force of gravity at work: In which direction is this force acting?

(d) forces that are lifting objects against the force of gravity: In which direction are they acting?

(e) a force produced by springiness or elasticity;

(f) a force that is slowing down the motion of an object;

(g) a force that is acting on an object in a downward direction.

Extension

The cartoon shows some actions that are dangerous. Identify actions that should not be done in real life.

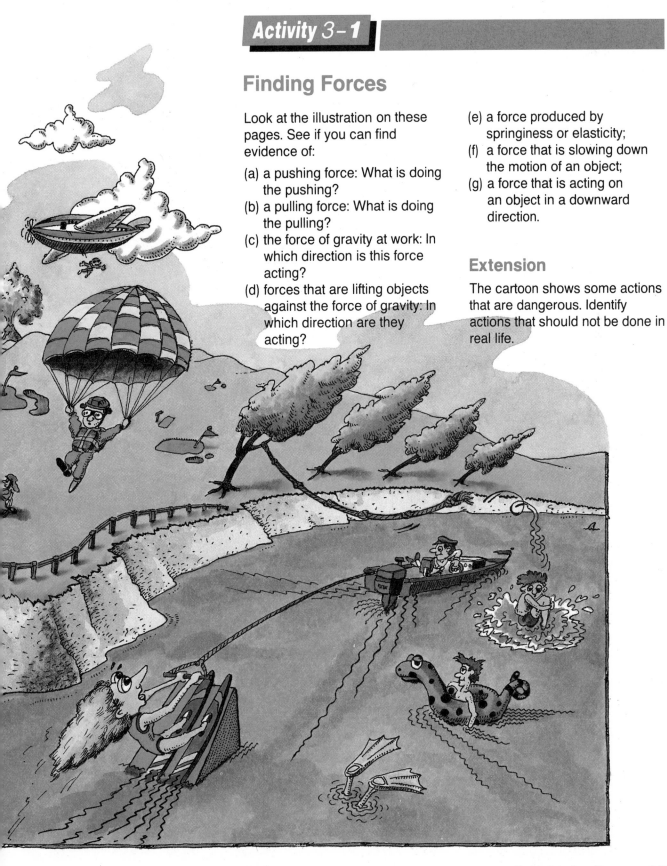

Experiencing Forces

Problem

What different forces are there?

Procedure

For each part of this Activity:
1. read the procedure;
2. do each step and observe what happens.
3. For each action you perform, note
   (a) whether the force is acting down, up, or sideways;
   (b) whether the force is a push or a pull.

PART A
Materials

water
pail or plastic container (3-5 L)
table tennis ball
rubber ball
styrofoam ball
steel ball

> **CAUTION: Do not use a glass container.**

Procedure

1. Drop a table tennis ball into a container of water.
2. Push the table tennis ball to the bottom of the container. Hold it for a short time and then let go.
3. Repeat the procedure with the rubber, styrofoam, and steel balls.

PART B
Materials

a penny
scissors
paper

Procedure

1. Cut out a piece of paper the size of a penny. Hold the paper in one hand and the penny in the other, both at waist height. Let them both drop at the same time.
2. Place the piece of paper on top of the penny and drop them together from waist height.

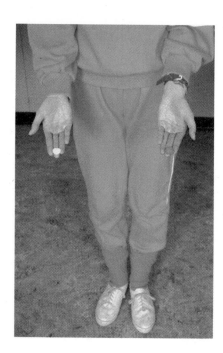

PART C
Materials

paperclip
magnet
index card
book
beaker
water

Procedure

1. Place a paperclip on a card. Hold a magnet under the card and move the magnet.
2. Try the same thing with a book between the paperclip and the magnet.
3. Place the paperclip in an empty beaker. Move the magnet around the outside of the beaker.
4. Fill the beaker with water and place the paperclip in it. Again move a magnet around the outside of the beaker.

PART D

Materials

paper
wool or fur
unused comb or plastic rod

Procedure

1. Cut up a few tiny pieces of paper. Push them around gently with a clean comb or plastic rod.
2. Rub the comb or rod on wool or fur. Again, put the comb near the pieces of paper.

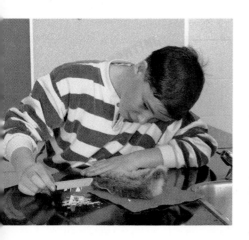

PART E

Materials

metre stick
rubber ball
styrofoam ball

Procedure

1. Place one end of a metre stick on the floor and drop a rubber ball from the height of 1 m. Use the metre stick to measure how high the ball bounces.
2. Repeat the procedure with a styrofoam ball.

Finding Out

1. Did any of your observations surprise you? If so, state what surprised you in each case.
2. In which part or parts was a force acting
   (a) downwards?
   (b) upwards?
   (c) sideways?

PART F

Materials

piece of rug or matting
piece of wood
2 bricks

Procedure

1. Lay a piece of rug or matting on the floor, and hold onto one end. Push a brick from one end of the rug to the other.
2. Push the brick across a piece of wood.
3. Push two bricks stacked on one another over one of these surfaces.

3. Which surface made pushing the brick more difficult?
4. Which part or parts demonstrated a force that was
   (a) slowing down motion?
   (b) speeding up motion?
5. How was pushing two bricks different from pushing one brick over the same surface?

Direction and Strength of Forces

The forces you experienced in Activity 3-2 were gravity, friction, magnetism, electrostatics, and buoyancy. Many of these forces affect you every day.

Forces may push or pull things upwards, downwards, sideways, or in any direction. Forces, by pushing and pulling, can even squeeze, bend, and twist things. Therefore, to understand how a force acts on an object, you need to know in which direction a force is acting.

Forces can be represented by arrows. The way the arrow points shows the direction in which the force is acting. The length of the arrow represents the strength of the force. For example, look at the series of illustrations of the girl kicking the soccer ball. They show the strength and direction of one force acting on the ball.

(a) A gentle kick.

(b) A hard kick.

Balanced and Unbalanced Forces

You have been thinking about one force acting on an object. But all objects have at least two forces acting on them—and sometimes more than two. For example, two forces are acting on the book in the illustration. A force you are very familiar with—gravity—is pulling down on the book. The force applied by the muscles in the hand is pushing up on the book.

(c) A kick in a new direction.

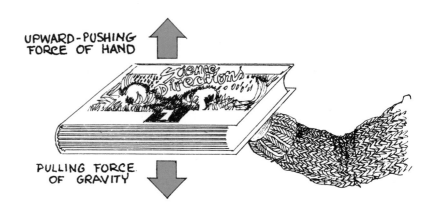

UPWARD-PUSHING FORCE OF HAND

PULLING FORCE OF GRAVITY

Look at the photograph of the weightlifter. Sketch him in your notebook. Draw arrows to show that his lifting force is greater than the force of gravity pulling the barbells down towards the Earth.

The two forces acting on the book in opposite directions are equal, and so the book does not move. When forces of equal strength act on an object in opposite directions, they are called **balanced forces**. For the book to move, either a new, third force must act on it, or one of the present two forces must become greater than the other. If gravity becomes the greater force acting on the book, the book will fall.

Now look at the illustration of the man pulling the boy on the toboggan. The long arrow represents the force that moves the toboggan forward when the man pulls on the rope. The short arrow represents the force of friction acting on the toboggan. **Friction** is a force that resists motion. It is acting to slow down the toboggan. These two forces, as the arrows show, are acting in opposite directions.

When one force acting on an object is greater than another force acting in the opposite direction, the forces are said to be **unbalanced**. Unbalanced forces will cause an object to begin to move. The direction of motion of the object depends on which force is greater. To start the toboggan moving forward, the man's pulling force on the toboggan must be greater than the force of friction.

You may find that using arrows to show both the strength and direction of all the forces acting on an object will help you to understand forces.

The arrows show the direction of the two forces. To start the toboggan moving, the pulling force must be greater than the force of friction, which resists movement.

Measuring Forces

How much force did it take to move the toboggan in the illustration? You are using your muscles all day long—do you know how much force your muscles use to open a drawer? Do they use less force to open a drawer than to lift this book?

We need to have a way of measuring the strength of forces. We use a clock or watch to measure time. We use a ruler to measure length. What kind of instrument could you use to measure the strength of forces?

In Unit Two you tested some ideas about structures and design. These ideas will help you to design a **meter**—an instrument used for measuring—to measure the amount of force needed to push or pull some common objects.

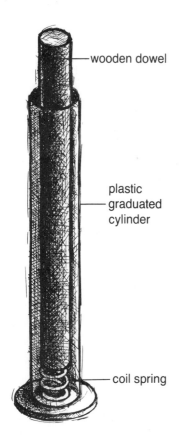

wooden dowel

plastic graduated cylinder

coil spring

Greg will have no difficulty pushing objects, but he will have problems pulling something. Can you make a suggestion to help him?

Designing and Building a Force Meter

1. To design and build a force meter, you need to think about
   (a) its function—how you are going to use the meter to measure the strength of forces that will move some objects in the classroom;
   (b) its structure and design—what different parts you will need to use in your force meter and how you will join them together.
2. Use this list to help you organize your thinking about the design of your force meter. In your notebook, jot down some ideas for each item on the list.
   (a) You will need to have a way to pull some things and push others: how will this affect your design?
   (b) When a force is tested, one part of the meter has to react to or resist that force. A springy or elastic material would resist the force being tested. By measuring the **resistance**, you can measure the strength of the force. What could you use that would stretch and return to its original length?
   (c) Can you find a way to make adjustments to your force meter to measure both strong and weak forces?
   (d) What could you use as a unit of measurement?
   (e) How could you mark a scale on your meter? (Read about scales in *Skillbuilder One*.
3. What could you learn from other people's designs? Have a look at what Greg and Debra did.

Greg's Meter

"There were a few things that I thought of measuring right away. I thought I'd test the force required to

- open the fridge door;
- lift a baseball;
- close a locker.

I started getting ideas for making my force meter by thinking about ways I could apply different amounts of force. I was clicking my pen nib up and down, so I thought of adapting the way my pen works. This is how I did it, using a wooden dowel and a plastic graduated cylinder. I figured I could use a weaker spring to measure small forces, and a stronger spring to measure larger forces."

Debra's Meter

"I like gadgets, so I thought of testing the force needed to push V.C.R. and T.V. buttons. I also wanted to test the tuning knob on my radio. As the knob turns right and left, I decided to put tape around it, then pull straight down on the tape.

I got some plastic pipe, a dowel, a hook, and some elastic. The illustration shows how I put these parts together.

As I wanted to measure weak forces, I thought washers would make good units. I decided to use five washers for one unit. I called my unit the debra (D) because I invented it. I marked zero (0) on the dowel at the bottom of the plastic pipe. Then I hung five washers on my force meter, and I marked the position of the bottom of the pipe on the dowel. I kept on adding washers, five at a time, and marking the position of the bottom of the pipe each time so that I had a scale marked on the dowel."

markings on dowel

metal washers

Debra's Scale

Debra invented her own unit to measure the strength of a force. Debra's unit of measurement was

five washers = one debra

She described how she marked a scale on her force meter, using her unit of measurement. Work out a scale using your own units of measurement.

Activity 3-3

Building Your Own Force Meter

1. Design and build your own force meter.
2. Next, test your meter by pushing and pulling on some small objects in the classroom. Does it work? Try it on some larger objects that need more force. The springy material should move a greater distance for large forces than it does for small forces. (You can be proud of yourself if your force meter works the first time. It won't be easy!)
3. If your meter did not work well, make some changes to the design and try again.
4. Test the strength of the force of 5, 10, and 20 of your units. You'll find it useful to have these measurements as standards for comparison.
5. Now measure the force needed to move 3 or 4 objects in the classroom. Record your measurements.

Finding Out

1. Did your force meter work well for the forces you chose to measure?
2. What changes might you want to make to the design of your force meter to improve its performance?

Finding Out More

3. Describe or draw the force meters in your class that are best at measuring
   (a) small forces;
   (b) large forces.
4. Debra (D) and Susan (S) compared their units and found that it took nearly 3 "S" units to make 1 "D" unit. Compare the units you used with those some of your classmates used.

Activity 3-4

Investigating the Strength of Forces

Problem

What strength of force is needed to move some common objects?

Procedure

1. In your notebook copy out a table similar to Table 1.
2. Look at the illustrations. Make an estimate of the strength of the force needed for the first action. (Make your estimate in the same units your force meter uses.)
3. Measure the strength of the force and record your result. Was your estimate close?
4. Select a word that you think best describes the direction of the force you measured.

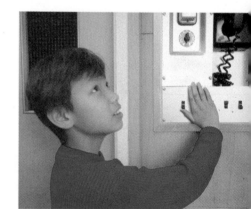

Investigating Forces

In the next Activity you will use the force meters you have designed to measure the strength of the force you use on some objects. Before you do the actual measuring, estimate the strength of each force. To help you make your estimates, think back to the classroom objects you tested in Activity 3-3. Would they be easier or more difficult to push or pull than the objects you will be working with in this Activity?

As you measure the strength of the force you are using on each object in Activity 3-4, think about the direction it is acting in, and how you would draw the arrow to represent the force.

Record this direction on your table. Some examples of words you might want to use here are "a pull out," "a push downwards," "a twist to the left."

5. Repeat steps 2 to 4 for each of the other actions illustrated. Your estimates will probably get more accurate as you go along.

Table 1

ACTION	NUMBER OF UNITS OF FORCE		DIRECTION OF FORCE
	ESTIMATED	MEASURED	
1. Opening your locker			
2. Turning on a light switch			
3. Opening a drawer of a filing cabinet			
4. Opening a sliding door			
5. Turning a doorknob			
6. Opening a refrigerator door			

Comparing Forces

You are going to test certain devices in your school to compare the force needed to operate each.

Procedure

1. Divide into groups. Each group will test a different device (for example, one group will test light switches, one group will test doorknobs).
2. As a group, make a list of four to six samples of the device in your school.
3. Each of you in the group measures every sample on the list, using his or her own force meter. Each of you records the measurements of every sample.

Finding Out

1. (a) Compare your measurements with those of the other members of your group. Which sample needed the least force to operate?
   (b) Did each member of your group agree?

Finding Out More

2. Yvonne, Tamil, and Fred decided to use their own force meters to test how much force would be needed to pull out drawers in kitchen cabinets and dressers in their homes. They each measured the force needed to pull out two drawers. Table 2 shows the measurements they made.
3. (a) Why do you think Yvonne's Tamil's, and Fred's measurements are so different?
   (b) Can you tell from the table which drawer requires the least force? Explain why or why not.
4. Explain why each member of your group had to measure the force needed to operate every device your group was testing.
5. How could the members of your group measure the forces needed to pull drawers at home and be able to decide which drawer needed the least force?

Table 2 *Force Required to Pull Out Drawers*

DRAWER	YVONNE'S MEASUREMENTS (Y UNITS)	TAMIL'S MEASUREMENTS (T UNITS)	FRED'S MEASUREMENTS (F UNITS)
1.	14		
2.	18		
3.		2	
4.		3	
5.			8
6.			10

Standard Units

The units you invented for your own force meter worked fine when you were measuring and comparing the strength of forces by yourself. But, as Yvonne, Tamil, and Fred found, you could not compare the measurements you made with the measurements someone else had made if you had not both used the same units to make the measurements. That is why people use systems of measurement that have **standard units**. Standard units always measure the same, wherever they are used. As you probably know, we use SI units of measurement in Canada. You are quite used to using the standard SI units for length, mass, and time. They are the metre (m), kilogram (kg), and second (s).

The standard SI unit for measuring force is the **newton** (N). The newton is named in honour of the great scientist, Sir Isaac Newton. Our knowledge about how forces cause motion is based on his ideas.

To get an idea of what a newton of force feels like, consider this. It takes about 1 N to hold up a mass of 100 g. That's about the same amount of force it would take to hold an apple or orange. It takes about 5 to 10 N to twist open an ordinary doorknob.

The next Activity will show you how to **calibrate** your force meter to measure in newtons. (To calibrate any meter is to make a scale for it.) Scientists throughout the world use and understand measurements of forces in newtons.

The force required to hold up a 100-g orange or apple is 1 N.

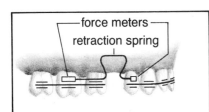

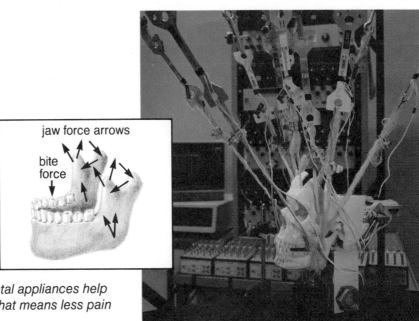

jaw force arrows

bite force

Precise measurements of forces in dental appliances help dentists straighten teeth more easily. That means less pain and more comfort for the patient.

The SI system was first used about 200 years ago. The full name of SI is the *Système internationale d'unités*. The name is in French because this system of measurement was designed and developed in France. The first standard unit of measurement developed in SI was the metre. That is why people often refer to the system as the metric system.

Using the Newton Scale

Problem

How do you calibrate your force meter in newtons?

Materials

the force meter(s) you designed
100 g masses
tape or piece of paper
marker

Procedure

1. Make sure the resistance in your force meter is operating consistently. If it is a spring or elastic, it should have been stretched and compressed at least 20 times. If it hasn't, give it a few extra stretches right now.
2. You need a clear place on your meter to mark a new scale of units in newtons. Try to design the scale so that you can still read your old units as well. It would be interesting to have both.
3. Start by marking 0 at the place on your new scale that shows that no force is being applied to the force meter.
4. Now hang a 100 g mass from your meter. Mark the new position on your scale as 1 N (one newton).
5. Now add a second 100 g mass and mark the place.
6. Repeat this procedure for as many masses of 100 g as your force meter will allow.
7. Now estimate and measure the force required to pick up several objects in the classroom.
8. Record your results in newtons.

Finding Out

1. What are the advantages of everyone using the same units to measure things?
2. Describe at least two other meters that measure in standard units. List the units of measurement they use and what each meter measures.

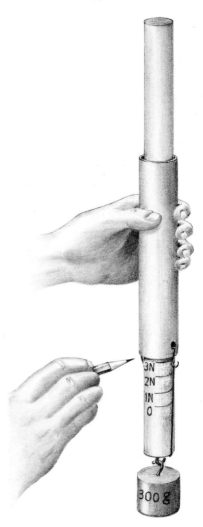

Checkpoint

1. Copy each of the six words below into your notebook. Match five of the words to the meanings given in (a) to (e). Write a meaning for the word that remains.

   force
   balanced
   unbalanced
   friction
   resistance
   standard

   (a) a force that resists motion
   (b) when one force acting on an object is greater than another force acting in the opposite direction
   (c) any unit that measures the same wherever it is used
   (d) when forces of equal strength act on an object in opposite directions
   (e) a force meter indicates the strength of a force when it measures this reaction to the force

2. Write this list of activities in your notebook. Beside each activity, state whether the force involved is a push or a pull, and describe the direction the force is acting in. If you think more than one force is involved, describe each force.
   (a) Pedalling a bike
   (b) Shooting a hockey puck
   (c) Using a teetertotter
   (d) Brushing your hair

3. Pick your favourite sport and list some of the movements that occur in it. Beside each movement write whether the force involved is a push or a pull.

4. Write four or five hints you would give to help someone build a good force meter.

5. How might you use your force meter to measure the force needed to pick up a pencil?

6. When you hang an object from a force meter, its spring or elastic band will stretch until the upward force is equal to the downward force.
   (a) What is that downward force called?
   (b) How can you tell when these two forces are balanced?
   (c) How can you tell when they are unbalanced?

7. (a) Draw a diagram of a diver or make a flip book of a diver. Have the diver run down the board, dive into the pool, swim to the surface, and get out of the pool.
   (b) Identify as many of the push or pull forces that the diver's muscles are exerting as you can. In which direction are these forces acting?

8. To find out how much force it will take to break sheets of metal, special force meters are used. Tom Villette of the University of Alberta is shown here preparing a 3 cm thick sheet of steel for the test. He found it took over 700 000 N of force to break the sheet steel.
   (a) List at least three products that are made from sheet steel.
   (b) State how you think Tom Villette's work would help the makers of each product.
   (c) If Tom Villette's force meter were designed for lifting, about how many oranges would it be able to lift?

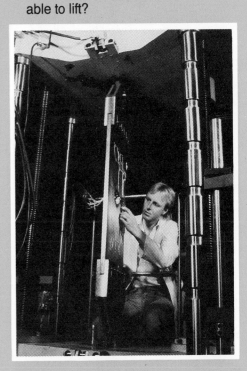

Gravity

Now that you have estimated and measured forces in the standard unit, the newton, you are ready to look closely at some specific forces. The first force you will investigate is one that you are familiar with because it pulls on you all the time—**gravity**. The Earth's gravity is the force that keeps your feet on the ground. When you drop a ball, or something falls, it is because gravity pulls the object towards Earth.

As you know, you need to consider two things about a force: direction and strength. On Earth, direction is easy: Earth's gravity always pulls down towards the centre of the Earth.

Gravity acts on everything on Earth: living things and non-living objects. All of these—your body, tiny micro-organisms, doorknobs, this book—are composed of **matter**. Matter is anything that takes up space and has **mass**. Gravity is the force that pulls anything that has mass towards any other thing that has mass.

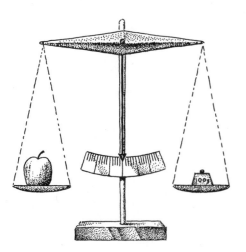

(a) Balance scales measure mass. The units for measuring mass are grams (g) and kilograms (kg).

(b) Spring scales measure weight. The unit for measuring weight is the newton (N).

Measuring Weight and Mass

You know that the standard unit for measuring force is the newton. When you measure the force of gravity on an object, you use a **spring scale** (force meter) to measure in newtons the **weight** of an object. When you measure the weight you are measuring the pull of gravity on the object.

The standard unit for measuring mass is the kilogram. When you measure mass, you use a **balance scale** to measure the mass of an object in kilograms. When you measure mass you are measuring the amount of matter there is in an object.

On a balance scale you compare (balance) the mass of the object you are measuring with masses calibrated in standard units (grams and kilograms). You put the object you are measuring on one pan of the balance scale, and a standard mass or masses on the other pan. When the two pans are at the same level, the mass of the object is exactly equal to the total of the standard masses in the other pan.

What Is Gravity?

Isaac Newton was the first person to explain gravity. In 1687 he published what he called his "laws of gravitation." Newton thought that all objects that have mass pull on other objects that have mass. He called the force that exerts this pull gravity. Newton then said that the strength of the force of gravity depended on two things. The first is *mass*. An object that has a large mass has a greater force of gravity than an object that has less mass. The Earth itself has a huge mass compared with an object on it. The Earth therefore has a very large force of gravity compared with the force of gravity of the objects on it. That is why the Earth's gravity holds objects on or close to the Earth's surface.

The second thing that affects the strength of the force of gravity is *distance*. The pull of gravity of one object on another is greatest when they are close together. The pull becomes weaker as they move farther apart.

Scientists still agree with and use Newton's ideas about gravity.

Weight and Mass

What is the difference between weight and mass? Mass—the amount of matter in an object—never varies. An object's mass is always the same wherever the object is on Earth, or wherever the object is in the universe.

But the weight of an object can and does vary, depending on where it is in the universe. This happens because

(a) different bodies in the universe have different masses;
(b) the mass of a body affects the strength of the force of gravity of that body; and
(c) the strength of the force of gravity affects the weight of an object.

On the surface of the Earth, a mass of 1 kg weighs just under 10 N (9.8 N to be exact). Think about a person with a mass of 60 kg on Earth. The weight of that person would be about 600 N. Now take that person to the Moon. The Moon's mass is only 1/6 that of the Earth. Therefore, the Moon's force of gravity is only 1/6 that of the Earth. When the person is weighed with a force meter on the Moon, she would weigh only about 100 N (1/6 of 600 N). But although she weighs only 1/6 as much on the Moon, her *mass* has not changed. Her body on the Moon still has the same 60 kg of matter.

An Interplanetary Holiday

This story is about how Shelly came to understand the difference between weight and mass.

It is the year 2222. Shelly is a student in Grade 7 and her science class is studying forces and motion. Her teacher returned from a holiday tour of several planets and mentioned something that puzzled Shelly. The teacher said that her *weight* was a lot greater on some of the planets than on others. This confused Shelly because her teacher had also said that things have the same *mass* wherever you are in the universe. Shelly decided she must really find the answer to this problem for herself. How could the teacher's weight change without her *mass* changing? Shelly talked her parents into lending her the keys to the Spacebug and their Plasticard to pay for fuel.

Shelly took the teacher's equivalent mass (68 kg) in silver with her. She grabbed a spring scale (force meter) and a balance scale and set off.

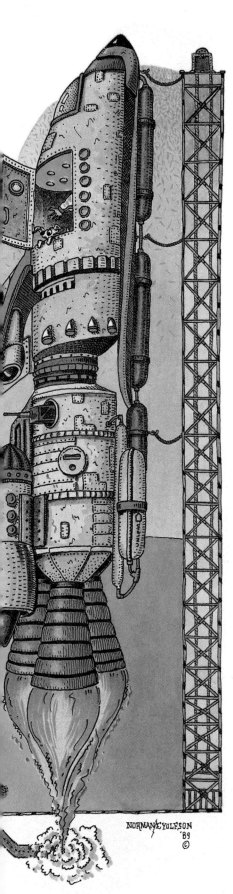

Shelly visited each of the inner planets: Mercury, Venus, Earth, and Mars. Each time she landed on a new planet she used the spring scale to measure the force of gravity in newtons on the 68 kg block of silver. She used the balance scale to make sure the mass remained at 68 kg on each planet. Here is a record of the data she collected.

Table 3 *Weight and Mass on Four Planets*

PLANET	WEIGHT (FORCE OF GRAVITY) OF THE SILVER BLOCK IN NEWTONS (N)	MASS OF SILVER BLOCK IN KILOGRAMS (kg)
Mercury	240	68
Venus	572	68
Earth	680	68
Mars	266	68

Shelly's teacher had been right. Weight varied from planet to planet but the mass remained the same.

Activity 3-7

The Force of Gravity

1. (a) Why do you think the weight of the silver block was different on different planets?
   (b) Which planet in Table 3 has the greatest mass?
2. Why didn't the mass, as measured on the balance scale, change from planet to planet?
3. (a) On which planet was the teacher's weight the greatest?
   (b) On which planet was the teacher's weight the least?
   (c) Which planet has the greatest force of gravity? How do you know?

(d) Which planet has the smallest force of gravity? How do you know?
4. If you could travel to any of these planets, which would you choose, and why, to
   (a) high jump?
   (b) play football?
5. Where would we be if gravity weren't holding things down? Write down all the benefits of gravity you can think of. Share your ideas with the rest of the class.

NORMAN EYOLFSON
'89
©

The astronauts in the American "Skylab" program were in space for several months. When they returned to Earth, they found that they were taller than when they had left. On Earth the bones in their spines were held tightly together by the Earth's gravity. In space, where the force of gravity was less, they had slight gaps between their bones. These gaps made them taller. (The astronauts returned to their normal height after several months back on Earth.)

Experiencing Less Gravity

We cannot escape the force of gravity. Astronauts in space, however, experience reduced gravity. Look at the diagram. As the astronaut's spacecraft travels farther from the Earth, the astronaut's weight will become less. If the spacecraft were 60 000 km from the Earth, the astronaut would weigh less than 10 N. This weight is so little that people often refer to this condition in space as "weightlessness."

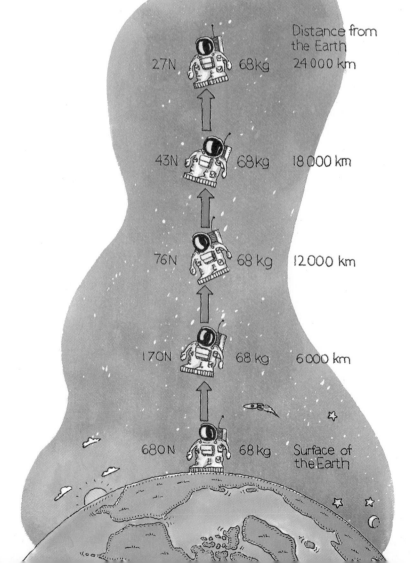

		Distance from the Earth
27 N	68 kg	24 000 km
43 N	68 kg	18 000 km
76 N	68 kg	12 000 km
170 N	68 kg	6 000 km
680 N	68 kg	Surface of the Earth

On the surface of the Earth, a 68 kg astronaut has a weight of 680 N. As the distance between the centre of the Earth and an astronaut increases, the force of gravity acting on the astronaut decreases. What happens to the mass of the astronaut?

The Moon has a much smaller mass than the Earth. Yet its gravity affects things here on Earth. Do some research to find out how the ocean's tides are produced. Prepare either a poster or a written report that shows why tides rise and fall more at some times than at other times.

These two pictures show the same spot on the Bay of Fundy in New Brunswick at high and low tide. The bay is shaped like a funnel. As the tide comes in, water rises higher here than anywhere else in the world.

Gravity of Other Objects

It is easy to understand that the gravity of the Earth pulls on you. But it is difficult to imagine that at the same time as the Earth is pulling on you, your own gravity is pulling on the Earth. Because your mass is so small by comparison with that of the Earth, you don't notice the effect of your gravity. Gravity also pulls you towards your desk and your desk towards you. This, too, is hard to imagine. Both your own mass and that of your desk are too small for these gravitational pulls to be noticeable.

Gravity Throughout the Universe

Newton thought that gravity acts between all objects throughout the universe. All the evidence that has been gathered supports Newton's idea. All the evidence to date supports Newton's other ideas about gravity, too. Here is a summary of his ideas about gravity:

1. The force of gravity is acting between *every* pair of objects in the universe.
2. The greater the mass of the objects, the greater the force of gravity between them.
3. The greater the distance between the objects, the smaller the force of gravity between them.

Think about an astronaut on a trip from the Earth to the Moon. As the astronaut moves away from the Earth, the force of gravity between the astronaut and the Earth becomes smaller. But gravity acts between all objects in the universe. As the astronaut moves towards the Moon, the force of gravity between the astronaut and the Moon becomes greater.

There is a special point, about nine-tenths of the way to the Moon, where the pull of the Earth's gravity is just balanced by the pull of the Moon's gravity. At that point the astronaut experiences a force of gravity of 0 N. Why do you think this special point is so much closer to the Moon than it is to the Earth?

Friction

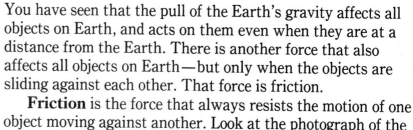

You have seen that the pull of the Earth's gravity affects all objects on Earth, and acts on them even when they are at a distance from the Earth. There is another force that also affects all objects on Earth—but only when the objects are sliding against each other. That force is friction.

Friction is the force that always resists the motion of one object moving against another. Look at the photograph of the downhill skier. On a steep slope, gravity is helping the skier to ski swiftly downhill. But friction between the skis and the snow acts in the opposite direction to slow the skier down.

Look back at the illustrations at the beginning of this Unit and find examples of the action of friction. In your notebook, sketch one example and label the two objects that are moving against one another. Use arrows to show the direction and strength of the forces acting on the two objects.

Measuring Friction

Friction must be overcome before an object will move. Suppose you want to move an exercise mat across a gym floor. You hook up a spring scale to the mat to see how much force it will take to start the mat moving. Suppose you find that it takes 100 N of force to begin to move the mat. That means that it took 100 N of force just to overcome the friction. The next Activity allows you to investigate the force of friction acting on an object.

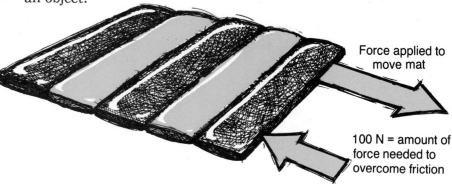

Force applied to move mat

100 N = amount of force needed to overcome friction

Friction must be overcome before an object begins moving. The friction force acting on the mat is 100 N. The force applied to move the mat must be more than 100 N.

Investigating Friction

Problem

What effect does friction have on the motion of an object?

Materials

2 bricks or blocks
spring scale (or force meter with a scale in newtons)
string
stack of paper or piece of cardboard
piece of wood (30 cm × 50 cm)
piece of linoleum (30 cm × 50 cm)

Procedure

1. Make a table similar to Table 4 to record your measurements.
2. Using the spring scale, pull the brick with a steady pull across the wood, as shown in the illustration.

Table 4

SURFACE	FORCE MEASURED IN NEWTONS (N)
Wood	
Linoleum	
Paper	

Table 5

SURFACE	PREDICTION OF FORCE NEEDED IN NEWTONS (N)	FORCE MEASURED IN NEWTONS (N)
Wood		
Linoleum		
Paper		

3. Record how much force is needed to keep the brick moving.
4. Pull the brick with the same steady pull across the linoleum and record the force you used.
5. Pull the brick with the same steady pull across the paper or cardboard and record the force you used.
6. Make another table, similar to Table 5. You are going to pull two bricks, stacked one on top of the other, across the three surfaces you have already tested.
7. Write in your table the force you predict will be needed to pull two bricks across each surface.
8. Pull the two bricks with a steady pull across each surface and record the force you used for each surface.

Finding Out

1. Compare the amount of force it took to pull one brick across each of the three surfaces.
2. What was the effect of doubling the number of bricks?
3. Now compare your results with your predictions.

Finding Out More

4. Use your results to help you answer the question below the illustration.

Drivers often put sandbags in the trunks of their cars when winter comes. Why do they do this?

Extension

5. Predict what would happen if the surfaces in Activity 3-8 had been wet. Try wetting each surface and repeating steps 2 to 5. How does wetting the surface affect your results?

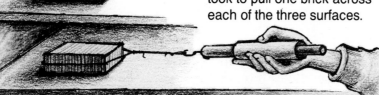
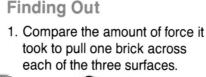

Friction **163**

Friction Summary

You have seen that:
- friction occurs when two objects move against each other
- the strength of the force of friction will vary depending on the roughness or smoothness of the surfaces that come into contact with each other
- the strength of the force of friction depends on the mass of the moving objects

Now take a closer look at these surfaces to find out why friction occurs.

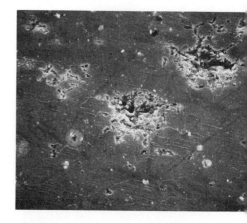

Under a microscope, even the smoothest surface looks rough. This is a view of the surface of a piece of polished steel magnified 250 times.

Activity 3-9

Studying Surfaces

Problem

Why does friction occur between surfaces moving against one another?

Materials

magnifying lens
stack of paper or piece of cardboard
brick
piece of linoleum
piece of wood

Procedure

1. Predict how the surface of brick, paper or cardboard, and wood will compare when observed closely.
2. Rub your finger over each of the three surfaces. Describe how each feels.
3. Look at each of the surfaces (brick, paper, and wood) under the magnifying lens. In your notebook describe and sketch what you see.

Why Friction Occurs

When you look at surfaces carefully, you can see that they are rough. The rougher the surfaces, the more they resist movement. Therefore, more force is needed to move objects with rough surfaces. Or, to put it another way, the rougher the surfaces, the stronger is the force of friction that resists movement.

The tracks on this bulldozer have huge ridges on them. Explain why a bulldozer has tracks like these.

Signs of Friction

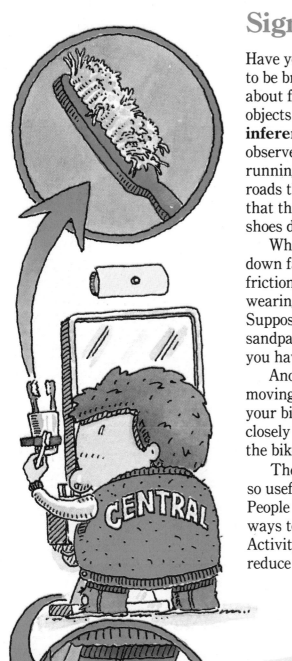

Have you ever had days when everything you go to use seems to be broken or worn out? Based on what you have found out about friction in this Topic, you could **infer** that each of the objects in the illustration has worn because of friction. An **inference** is a possible explanation for something you have observed. For example, you notice that the soles of your running shoes wear down faster when you run on rough gravel roads than when you run on smooth paved roads. You infer that the rough gravel surface wears the soles of your running shoes down faster than the smooth paved road.

When two rough surfaces rub against each other, they wear down faster than two smooth surfaces. Wearing is an effect of friction—one that is often unwanted. However, this same wearing action of friction can sometimes be put to good use. Suppose you want to paint some wooden shelves. You use sandpaper to wear down the rough surfaces of the wood so that you have a smooth surface to paint.

Another effect of friction—the ability to slow down or stop moving objects—can be a lifesaver. Think about the brakes on your bike or the family car. At the next opportunity, look closely at how the brakes on a bike use friction to slow down the bike or make it stop.

The action of friction in slowing down moving things is not so useful when friction acts on the moving parts of machinery. People in industry have spent much time and effort finding ways to reduce friction when it is not wanted. In this next Activity you will test some substances that might help to reduce friction.

Reducing Friction

Problem

What are some ways to reduce friction?

Materials

spring scale
brick
piece of wood
fine sandpaper (150 grit)
coarse sandpaper (80 grit)
liquid detergent or liquid soap
marbles in lids of jars
magnifying lens
wax paper
string
7 or 8 straws or dowel pins (laid side by side in row)

Procedure

Use the items listed here, and your observations in Activities 3-8 and 3-9, to find at least three possible methods to reduce friction between the brick and the wood.
For each method you invent:

• predict the amount of force required to pull the brick;
• inspect the surface with your magnifying lens and rub it with your finger;
• pull the brick across the surface with a spring scale and record your results.

Finding Out

1. Explain why some of your methods were more successful than others in reducing friction.

Extension

2. Hard surfaces that move against each other in a machine are often protected by **bearings**.
(a) Look at the illustration of the bearings in the skateboard. What means of reducing friction do bearings use?
(b) Look at a bike and find all the bearings.

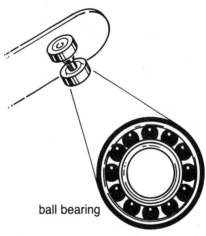

ball bearing

The picture shows a close-up of the bearings in the wheels of a skateboard.

If you were to use graphite or silicone on this bicycle to reduce friction, which spots would you spray? What did you use for this purpose in Activity 3-10?

(c) How are the bearings different in some places on the bike than in others?
(d) Why might they be different?
(e) In your notebook, sketch the different kinds of bearings, and describe the function each performs.
3. Look at the illustrations and answer the questions.

Bobsled teams spend hours smoothing and polishing the runners of their sleds. What did you do in the Activity that was similar to what the bobsled teams do?

Other Effects of Friction

Complex machines, like cars, have many moving surfaces that rub against each other. **Lubricants** are used in machines to reduce friction and wear on the moving parts. A lubricant is a substance that reduces friction between moving surfaces. The liquids you used in Activity 3-10 are examples of substances that can act as lubricants. Other examples are oils and greases.

Lubricants also help to reduce one other effect of friction, which you will explore in the next Activity. And like other effects of friction, this effect can be helpful or harmful to us, depending on the circumstances.

Activity 3-11

Another Effect of Friction

Problem

What other effect does friction have?

Materials

brick
large nail

Procedure

1. Rub the end of the nail vigorously against a brick for 10 s.

> **CAUTION:** Do not rub the nail any longer than 10 s.

2. Quickly (but carefully) touch the end of the nail to your other hand.

Finding Out

1. What two things have happened to the end of the nail?
2. What would happen if you increased the force of the nail against the brick by pressing harder? If you decreased the force? Try it and record your observations.

Finding Out More

3. Think of some situations where this effect of friction could be useful.
4. Think of some occasions when it might be harmful.

Extension

5. Think of some instances where friction has caused a disaster. Bring articles or pictures to class to illustrate your examples.

More about Friction

As you know, friction occurs when two surfaces move against each other. You have tested a variety of objects moving against solid surfaces such as wood and linoleum. It may surprise you to realize that friction also acts in liquids, such as water, and gases, such as air.

The following thought experiment will help you recall some of your own experiences with friction in air and water.

Activity 3-12

Friction in Water and Air

1. Imagine that it is a beautiful, warm, and calm day. You set out on a bicycle ride.
   (a) What do you feel against your skin, hair, and clothes as you pedal along?
   (b) How does what you feel change as you move faster?
   (c) What do you infer is causing these effects?
2. Imagine you are at the beach with a friend. You decide to have a race.

(a) How does running in the water compare with running on land?
(b) What two surfaces are moving against each other as you move in water?
(c) Why is it harder to work against friction in water than friction in air?

L. SENIN

Friction: Harmful *and* Helpful?

As you can see, friction is a powerful force that resists motion between surfaces, whether they are solid, liquid, or gas. The friction caused by air (air resistance) is so small at low speeds that we often disregard it. As you get to higher speeds though, it could be a fatal mistake to ignore air resistance. The story gives an example of a situation where managing friction from the air becomes a matter of life or death.

A Shuttle Story

Read the following story and be prepared to answer the questions.

For the past week the shuttle has been orbiting about 200 km above the Earth's surface. At this distance from the Earth there is very little air and therefore very little friction. Now it is time for the shuttle to leave space and return to Earth. The engines are fired briefly to slow the craft down to 28 000 km/h. At this speed the shuttle leaves its orbit and falls steadily towards the Earth.

When the shuttle has fallen to a distance of 130 km above the surface, the astronauts can hear air beginning to brush the sides of the shuttle. The temperature on the outer surface of the shuttle begins to rise because of the friction from the air at this high speed. But, as planned, the special tiles on the outside of the shuttle protect the shuttle and the astronauts inside it from an increase in temperature that could be fatal.

As the shuttle continues to fall, the friction between the air and the shuttle acts as a brake to slow the craft down to 2000 km/h in a matter of minutes. As the speed drops, the friction lessens, and the tiles on the outside of the shuttle begin to cool. Finally, the shuttle glides in zigzags to lose speed and lands safely at a speed of around 300 km/h.

Finding Out

1. What problem does friction cause the spacecraft?
2. What benefits does it provide?
3. Describe ways in which the return to Earth would have been different if there were no friction.

Friction in Everyday Life

This story has shown an extreme example of how friction is both helpful and harmful. Is friction both helpful and harmful to you? Think about it next time you are riding your bike, using power tools, ice skating, swimming, or sliding into first base.

1. Complete the word game to find the name of something that takes up space and has mass.

(a) ☐ ■ ■ ■
(b) ■ ■ ☐ ■ ■
(c) ■ ■ ■ ☐
(d) ■ ■ ■ ☐ ■
(e)   ■ ☐ ■ ■ ■ ■
(f)   ■ ☐ ■ ■ ■ ■ ■

(a) the amount of matter in an object
(b) what you make when you calibrate
(c) the metre and the kilogram are examples of this
(d) the standard unit for measuring force
(e) the force of gravity on a mass
(f) the name of the force that pulls objects towards one another

2. This picture shows Hans standing on top of Mt. Assiniboine, and by the ocean. Will his mass be different on top of the mountain than it is by the sea? Why?

3. Draw a table with one column for mass, and another column for weight. Place each of the following terms under the appropriate column.
(a) kilogram
(b) newton
(c) spring scale
(d) balance scale
(e) force of gravity
(f) the same on every planet
(g) different on every planet

4. Imagine that for one day the force of gravity on Earth was as low as that on the Moon. Describe how your life would change. What things would be more fun? What things would be difficult, even impossible?

5. Fuel economy in space is incredible! Space probes can travel millions of kilometres on only one litre of fuel. Explain why.

6. What do racing cyclists do to reduce the frictional force acting against them as they ride their bikes?

7. Explain why the soles of climbing boots have ridges.

8. In your notebook draw a table with the headings shown below, and fill it in.

THREE RESULTS OF FRICTION	AN EXAMPLE WHERE THIS RESULT IS USEFUL	AN EXAMPLE WHERE THIS RESULT IS NOT USEFUL

Other Forces in Action

Friction is not the only force that results when surfaces of two different kinds of materials rub against one another. The rubbing may produce another force you are quite familiar with, too. You may call this force "static." The electricity produced from the rubbing of two surfaces against each other results in a force that scientists call **electrostatic force**.

Think back to your own experiences when "static" made your clothes cling to you, or pulled your hair towards your brush as you brushed your hair. Have you ever got a shock from a doorknob or other metal object? Your clothes cling because of an electrostatic force that pulls the cloth against your skin. Your hair is pulled or attracted to your brush by this same force. And the shock you received from the doorknob is also because of this force.

The following Activity will give you a chance to investigate electrostatic forces.

Performing Paper

Problem

Can an electrostatic force move objects?

Materials

an unused plastic comb, glass rod, or clear plastic pen sheath
fur or wool
strip of paper (2 cm × 8 cm)

Procedure

1. Make the paper into a loop (use a small amount of glue). Make the paper loop as round as possible so that it will roll easily on a smooth surface.

2. Rub the clean comb or plastic rod vigorously with the wool or fur.
3. Predict what will happen when you bring the comb or rod nearer and nearer to the paper loop. Test it. Record the approximate distance at which something happens to the loop.
4. Predict what will happen if, without touching the loop, you try to move it around the table top by using the comb or rod. Test it.
5. Predict in which position the comb or rod will work best to move the loop. Test these positions.
6. Repeat steps 2 and 3 using a wet comb or rod.

Finding Out

1. Which force did you use to make the electrostatic force on the comb or rod?
2. Did the electrostatic force push or pull the loop?
3. What was the effect of adding water to the plastic comb or rod?

Finding Out More

4. What effect did distance from the object (loop) have on the strength of the electrostatic force?

Extension

5. Set up an "obstacle course" or a maze through which you can direct your paper loops. See who can get the best time. Analyse the different ways of using the comb or rod to determine why some work better than others.
6. Rub two plastic straws with wool or fur. Lay one straw on a table and bring the other straw near it. What happens? Why? A reference book could help you find an explanation.
7. Try blowing some soap bubbles. Rub your comb or rod with wool or fur and bring it near to them. What happens? Explain why.

Getting a Charge

When an object like a plastic comb or rod exerts a push or pull that is due to electrostatic forces, we say it is electrostatically charged. These electrostatic charges can leap from a charged object to create a spark. Perhaps the most spectacular demonstration of electrostatic sparking is lightning. Water droplets and air moving in a cloud produce an electrostatic charge in part of the cloud. When this charge leaps to another part of the cloud or the ground, it produces lightning.

Magnetism

Magnetism is a force that pushes or pulls objects in the invisible field around a magnet. Every time you close your refrigerator door you use this force as well as that of your own muscles. (Next time you close your refrigerator door at home, close it slowly to within 1 cm. You can feel the pull that causes the door to close the rest of the way.) You also use magnetic force to play tapes on a tape recorder, to ring doorbells, and in many other ways each day of your life.

One end of a magnet is called the North pole. The other end is called the South pole. The North pole of one magnet and the South pole of another magnet pull towards each other. If their North poles are placed near each other, they push away from each other. So will the two South poles push away from each other.

The following Activities will give you a chance to learn more about magnetism and the fields of force that surround magnets.

Magnetic Manipulations

Problem A

What does a magnetic force field look like?

Materials

2 magnets
iron filings
transparency sheet or glass
(paper will do if these are not available)

Procedure

1. Cover one magnet with the transparency.
2. Carefully sprinkle iron filings on the transparency and tap it gently.
3. Draw in your notebook the pattern of filings you see.
4. Repeat this procedure using two magnets.
   (a) First place them about 5 cm apart with the North pole of one and the South pole of the other magnet facing each other.
   (b) Then place them the same distance apart with the two North poles facing each other.

Problem B

How can I make a magnet?

Materials

steel rod, metal knitting needle, or large nail
strong magnet
test items (paperclips, small nails, iron filings)

Procedure

1. Stroke the rod or needle with the magnet. After each stroke, lift the magnet well above the rod and return to the starting point. Be sure always to stroke in the same direction.
2. Repeat 40 times.
3. Now test your new magnet by trying to pick up some of the test items. Record your observations of its strength.

Problem C

Can magnetic forces act over a distance?

Materials

your new magnet or a classroom magnet
paperclip
string (about 30 cm long)
ruler
lab stand and clamp

Procedure

1. Tie a paperclip to one end of the string.
2. Tape the other end of the string to your table.
3. Hold the magnet straight up and down with a clamp.
4. Bring the paperclip close to the magnet so that the paperclip is held up by the magnet but does not touch it.
5. Slowly pull the magnet away from the paperclip until the paperclip drops.
6. How far can the paperclip be from the magnet before the paperclip falls? Record your results.

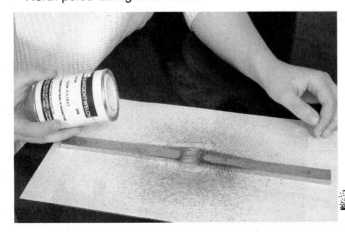

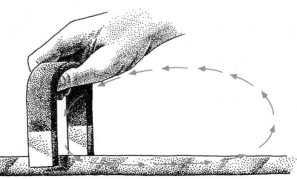

Problem D

Can a magnetic force act on different metals?

Materials

magnet
copper penny
nickel
steel nail
aluminum nail (roofing nail)
paperclips

Procedure

1. Touch the magnet to the steel nail.
2. Bring the nail close to the paperclips and observe what happens.
3. Repeat the procedure with the aluminum nail, the penny, and the nickel.

Finding Out

1. In Part A you saw a pattern made by iron filings. The filings tend to collect in lines where the magnetic force is strongest. Most of the filings are collected near the ends or poles of the magnet. This tells you that the strength of the magnetic force is strongest there. How would you explain the two different patterns that formed in Step 4?
2. Part B demonstrated that one magnet can be used to make another. The makers of magnetic compasses use this procedure to magnetize the compass needle. Can you think of a way you could easily make your own compass?
3. Part C showed that an object does not have to touch a magnet to be held by its force. Where might you use a magnetic force that does not touch the object it is acting on?
4. In Part D what did you learn about magnetic forces being able to act on different metals?

Extension

5. The force of magnetism acts on a variety of materials containing iron, such as nails, filings, ball bearings, and bolts, causing them to stick together. Try creating your own magnetic sculpture by arranging a variety of such objects around a magnet.

Buoyancy

Have you ever tried lifting a friend in a swimming pool? Did it seem that your friend had suddenly become lighter? Objects seem lighter, and therefore *seem* to have less mass, when they are immersed in water. But your friend's mass did not actually change in water. What did change was your friend's weight, because the forces acting on your friend changed. The water exerts an upward push. This upward force is called **buoyant force**, or **buoyancy**. Fluids such as water exert a buoyant force on objects.

A person's mass seems less in the pool, but is it really?

The Force of Buoyancy

Problem

What is the strength of a buoyant force acting on an object in water?

Materials

a large plastic beaker or bowl
a spring scale
string
a stone
various classroom objects

Procedure

1. Suspend the stone from the spring scale and weigh it in newtons.
2. Lower the stone into the water and weigh it again. (Do not let the stone touch the bottom of the bowl.)
3. In the illustration,
   • the stone weighs 5 N in air
   • the stone weighs 3 N in water
   • the difference is   2 N

Therefore the force of buoyancy on the stone is 2 N. Do this calculation for the stone you measured.
4. Repeat this procedure with other objects.

Finding Out

1. What is the main force acting on the stone before it is put in the water?
2. In which direction is this force acting?
3. What forces are acting on the stone when it is immersed in the water?
4. In which direction is each of these forces acting?
5. Draw the stone immersed in water with force arrows showing the two forces acting on it.

Extension

6. Make a boat using modelling clay. Design your boat so that it will hold as many marbles as possible and still float.

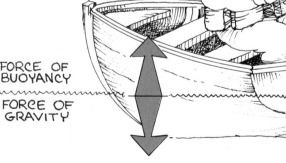

Floating and Sinking

Some objects float in water. When an object is floating in a liquid such as water, the forces of buoyancy and gravity acting on it are equal. An object will sink if the force of gravity acting on it is stronger than the upward buoyant force.

FORCE OF BUOYANCY

FORCE OF GRAVITY

Collect several objects made of different materials. Predict whether or not they will float. Place them in your pail to check your predictions. Some interesting objects to try are oranges or apples, ice cubes, corks, paperclips, hard-boiled eggs, and erasers.

Activity 3-17

Buoyant Forces of Liquids: Demonstration

Problem

Do other liquids have the same buoyant force as fresh water?

Materials

1 uncooked egg
4 glasses
warm water
salt
rubbing alcohol
cooking oil

Procedure

1. Half fill the first glass with warm water and carefully place the egg in it. Record what happens.
2. Half fill the second glass with warm water and dissolve some salt in it. Again, gently place the egg in it. Record what happens.
3. Half fill the third glass with rubbing alcohol. Carefully place the egg in it. Record what happens.

4. Half fill the fourth glass with cooking oil. Carefully place the egg in it. Record what happens.

Finding Out

1. Which liquid exerted the largest upward push on the egg?

Finding Out More

2. Where do you think it would be easier to learn to swim, in the salty water of the sea or in a freshwater lake? Why?

Extension

3. A buoyant force also acts on objects that float in the air. Helium balloons, for example, rise in air. Can you think of other examples of buoyancy in the air?

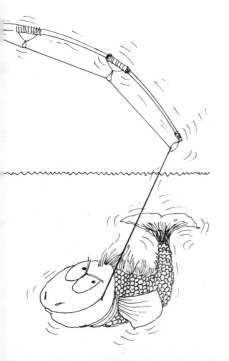

Science and Technology in Society

Girls and Boys, Ladies and Gentlemen, presenting... Science in the Circus!

Imagine yourself poised on the end of a teeterboard. Any second now you will be hurled upward and backward through the air. You hope to land squarely on the shoulders of another circus performer. You've done this act many times before, but you still wonder: "Will I make it?"

You've done it! Quickly you grab your partner's hands. Safe at last.

Just how did you do it? Two other circus performers jumped on the other end of the teeterboard. They applied a strong downward force, giving an upward thrust to your end of the teeterboard, and away you flew.

Where might you have ended up if three or four acrobats had jumped on the teeterboard and increased the downward force? According to André Simard, the Technical Director of Acrobatics at the National Circus School, a great deal of experimentation and calculation is needed to answer such a question. In fact, he says, circuses are increasingly relying on scientific ideas to work out how to use forces in their acts.

Circus acts work with the forces of nature to create dazzling displays. Yet many acts—tightrope walking, aerial acrobatics—are particularly thrilling and terrifying because they seem to challenge and defy these forces. As one acrobat put it, "A rock or a human being flying through the air must both come down. Gravity is the main thing we are fighting here."

Friction, however, is an aerial acrobat's friend. A good grip on the trapeze, on the ropes, and on the partner is essential. The smooth metal trapeze bar is wrapped in special cloth tape to provide a firm grip. Before the show, the acrobats apply resin or chalk to their hands to increase their grip. They may use fitted leather hand grips as well. The acrobats often wrap their wrists with gauze for those terrifying moments when they must grab each other's wrists in midair.

Circus costumes are designed with great care. Imagine an aerial acrobat trying to catch and hold someone wearing a slippery silk shirt!

Circus performers, like all athletes, choose their footwear carefully. They may wear gymnastic slippers, but sometimes they use rubber-soled Tae Kwon Do shoes for extra grip and support. Rubber-soled shoes provide almost twice the frictional force of leather-soled gym slippers.

Imagine a tightrope act on the Moon, where there is only 1/6 the gravity of Earth. Would the performance seem as exciting as here on Earth?

Think About It

1. Look at the circus acts and imagine which ones would be easier or more difficult with
   (a) more gravity than on Earth;
   (b) less gravity than on Earth.
2. Would a juggler be able to perform on a space vehicle where there is little gravity?

A Close Look at Motion

In 1687 Isaac Newton published his ideas about gravity and **motion** in a book. These ideas are still, more than 300 years later, supported by the evidence. The only exceptions to his ideas are objects travelling very, very fast—near the speed of light. (Three hundred million metres a second—now that's fast!)

Newton's ideas about motion can be summarized in his Three Laws of Motion. The First Law states in part that *an object at rest (in other words, not moving) tends to stay at rest unless a force is applied to it.*

A MOVEMENT FROZEN!
AN INTRODUCTION TO ISAAC NEWTON'S FIRST LAW OF MOTION.

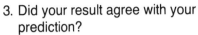

The Penny and Card Trick

Problem

How does the penny and card trick support Newton's idea about objects at rest?

Materials

index card
penny
water glass or beaker

Procedure

1. Lay the card on the desk and put the penny on it.
2. Predict what might happen when you pull the card sharply out from under the penny.
3. Try it and see what happens.

Finding Out

1. (a) Did either the card or the penny move? In which direction?
   (b) Did either the card or the penny stay still?
2. (a) Did you apply a force to the penny?
   (b) Did you apply a force to the card?
3. Did your result agree with your prediction?
4. How does this trick support Newton's ideas about objects at rest?

Extension

5. Place the penny on the card and put both on the glass. Predict what will happen when you flick the card rapidly sideways with your finger.
6. Flick the card and see what actually happens.
7. (a) Did the card move? If so, in which direction did it move?
   (b) Did the penny move? If so, in which direction did it move?
   (c) Did a force act on the card? If so, what force was it?
   (d) Did a force act on the penny? If so, what force was it?
8. Explain how this trick supports Newton's ideas about objects at rest.

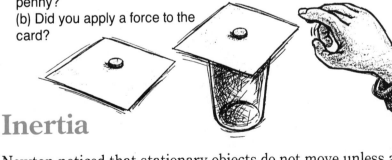

DID YOU KNOCK OVER? WHY DID STAY STILL, UNTIL HIT ME?

... GOODBYE ... SIR?

ONE OF NEWTON'S IMPORTANT IDEAS IS THAT OBJECTS STAY AT REST UNTIL A FORCE IS APPLIED TO THEM — ONLY THEN DO THEY MOVE.

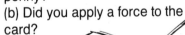

Inertia

Newton noticed that stationary objects do not move unless a force acts upon them. This resistance to change is called **inertia**. Newton also noticed that objects already moving continue to move at the same speed and in the same direction, unless an unbalanced force acts on them. This resistance to change is also called **inertia**.

Stop and Start

1. Look at illustration A. Tina is embarrassed. This is the first time she has had to stand on the bus. When the bus started, she fell backwards into the lady behind her.
   (a) Explain why Tina fell backwards.
   (b) What would you advise Tina to do next time the bus starts?

A

2. Now look at illustration B. The bus driver has braked sharply, and the bus has stopped suddenly.
   (a) What happens to Tina when the bus stops?
   (b) Explain why this happens.

B

3. Look at illustration C. Tyler has to deliver this large pile of newspapers. As he starts out, the papers fall off the back of the wagon.
   (a) Explain why the papers fall.
   (b) What would you advise Tyler to do next time he starts out?

C

4. Look at illustration D. Tyler has put the papers back on the wagon. He starts off again. The wagon wheel hits a rock and the papers fall off the front of the wagon. What advice can you give Tyler now?

D

5. Use what you know about the inertia of moving objects to explain why it is always a good idea to do up your seat belt.

Mass and Inertia

Another of Newton's ideas about motion is that an object that has a large mass has more inertia than an object that has a smaller mass. The illustration shows that an object that has a large mass is harder to start moving, and harder to stop moving than an object that has a smaller mass.

A has more mass and greater inertia at rest than C.
B has more mass and greater inertia in motion than D.

Activity 3-20

Demonstrating Mass and Inertia

Problem
How does mass affect inertia?

Materials
table tennis ball
golf ball

Procedure
1. Look at the illustration and answer the question.
2. Predict what will happen when you place a golf ball on a smooth floor and roll a table tennis ball into it. Try it.

3. Now predict what will happen when you roll the golf ball into the table tennis ball. Try it.

Finding Out
Explain your results in terms of what you have learned about mass and inertia.

The truck and the car are travelling at the same speed. They are both going to brake at the line. Which do you think will require more breaking force to bring it to a stop? Why?

Action and Reaction

So far you have looked at one or more forces acting on a single object. Now think about two objects: one that exerts a force and one that receives a force. Newton observed that every time an object exerts a force on another object, the receiving object exerts an identical force in the opposite direction. Newton stated that *for every action, there is an equal and opposite reaction.*

The illustration shows a familiar example of **action and reaction.** When you walk, you push backward against the ground (action) and the ground pushes forward against your foot (reaction). If you do not believe that the ground pushes you forward, why do you think it is so difficult to walk in deep snow or sand? Imagine what it would be like walking on marbles. Try the following Activity to observe this action and reaction for yourself.

Activity 3-21

Equal and Opposite Reactions

Problem
What is the action and what is the reaction as a car moves?

Materials
sheet of stiff cardboard
(10 cm × 20 cm)
marbles
wind-up or battery-powered toy
car

Procedure
1. Put the marbles close together on a smooth surface, and carefully lay the cardboard on top of them.

2. Predict what will happen if you put the moving car on the cardboard.
3. Wind up (or turn on) the toy car and set it carefully on the cardboard.
4. Observe and record your results.

Finding Out
1. In what direction do the wheels push on the cardboard? How can you tell?
2. How does a car on a road move forward? Explain in terms of action and reaction.

More Examples of Action and Reaction

Look at the pictures of the student on the skateboard. As the student throws the ball forward, the skateboard (with the student on board) begins to move backward. The forward push the student gave the ball supplied the action force on the ball. The backward movement of the student and skateboard is evidence of the reaction force of the ball on the student. The reaction force is equal in strength to the action force, but the mass of the ball is much less than the combined mass of the student and the skateboard. Because the ball's mass is so much less, it moves much farther than the student on the skateboard.

According to Newton's Law of Action-Reaction (as it is sometimes called), forces come in pairs. Each force in the pair acts on a different object. Using this knowledge, scientists have been able to get rockets off the ground. The force of the rocket pushing on gases spewing from it is the action. The reaction force of the gases pushes the rocket upward. Do the following brief Activity that serves as a model for rocket movement.

Activity 3-22

Investigating Action-Reaction Forces: Demonstration

Problem

How can a balloon serve as a model of action-reaction forces in rockets?

Materials

2 small (5 cm) binder rings
1 long balloon
string
tape
2 chairs
spring clip

Procedure

1. Blow up the balloon and close the end with a spring clip.
2. Tape the two rings onto the balloon and run the string through them.
3. Arrange the chairs about 4 m apart and tie the string between them.
4. Pull the balloon to one end of the string and release the clip.

Finding Out

1. (a) What is the action force?
   (b) What is the reaction force?
2. In which direction is each force acting?
3. Explain how this Activity serves as a model of action and reaction forces in a rocket.

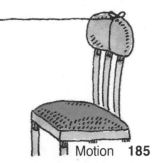

Satellites in Motion

The Moon is a satellite of the Earth. It has been orbiting the Earth for millions and millions of years. The first satellite made by people is much more recent. It is *Sputnik 1*, and was put into orbit by the Russians in 1957. Now there are hundreds of satellites moving around the Earth in various orbits.

A **satellite** is any object in space that orbits a larger object. How does a satellite get into orbit? As you may have seen in pictures or on television, a satellite is placed on a rocket. The rocket's engines spew out gases that push the rocket upwards with tremendous force (called thrust). At a certain height above the Earth, the satellite is sent off by a blast from the rocket's engines. The satellite then orbits in a curved path without further help from the rocket.

Why does a satellite follow a curved path? According to Newton's first law of motion, an object moving in a straight line will continue to move in a straight line unless a force acts on it. Why doesn't the satellite continue in a straight line after being sent off by the rocket? The downward pull of gravity causes the satellite to follow a curved path. If it weren't for gravity, the inertia of the satellite in motion would keep it moving in a straight line that would carry it away from the Earth.

Since gravity does pull on the satellite, you might ask why a satellite stays up? Why doesn't it fall to Earth? The answer lies in the satellite's forward motion. All along the path of its orbit, gravity keeps pulling the satellite down towards the Earth. But the satellite's forward motion keeps it from falling to Earth. The satellite, therefore, follows a curved path around the Earth because of two things—its forward motion and the downward pull of gravity. In the next Activity you will construct a model of the flight path of a satellite in orbit.

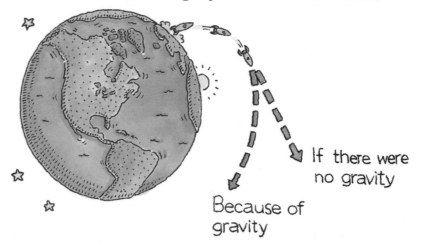

If there were no gravity

Because of gravity

Balanced Forces in Orbit: Demonstration

Problem

What forces are acting on an object in orbit?

Materials

15 cm long copper or PVC tubing with ends smoothed carefully
100 g of metal washers
fishing line
one-hole rubber stopper
2 paperclips

Procedure

The demonstrator will carry out the following steps.

1. Thread the string through the tubing, leaving 50 to 60 cm hanging out at each end.
2. Tie the rubber stopper to one end and a paperclip to the other end of the string. Bend the paperclip to form a hook and hang 5 washers from it.
3. Tape the other paperclip about 30 cm above the first one (this will act as an indicator of the string's movement).

> **CAUTION:** Everyone should be at least 3 m from the model when it is being demonstrated.

4. Twirl the rubber stopper above head height, holding the tubing securely.
5. Add more and more washers to the hook while twirling the stopper. Try to keep the paperclip indicator at the same height by adjusting the twirling speed.

Finding Out

1. What object in this model represents the satellite?
2. What object in this model represents the Earth?
3. When more washers are added, do you need to increase or slow the twirling speed to keep the indicator at the same height?

4. What force was pulling the "satellite" towards Earth?
5. What force pushed the "satellite" away from Earth?
6. In which direction was each force acting?

Finding Out More

7. (a) What would happen if the force in 4. stopped acting?
   (b) What would happen if the force in 5. stopped acting?
   (c) What would happen if you shortened or lengthened the string (the height of the orbit)?
8. How is this model different from the orbiting of a real satellite?

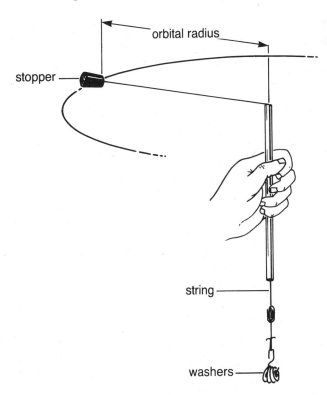

stopper

orbital radius

string

washers

Satellites for Canada

Canada is the second largest country in the world. One problem for such a large country with a small population has been to find ways for people to communicate with one another. Radio and TV signals can be sent long distances, but they can be blocked by mountains or disrupted by the weather.

Canadian scientists researched ways to overcome these problems. As a result of their studies, they have become world leaders in the design, construction, and operation of communication satellites. Through their efforts, we have become quite accustomed to watching international programs, news, and sports events live on TV "by satellite."

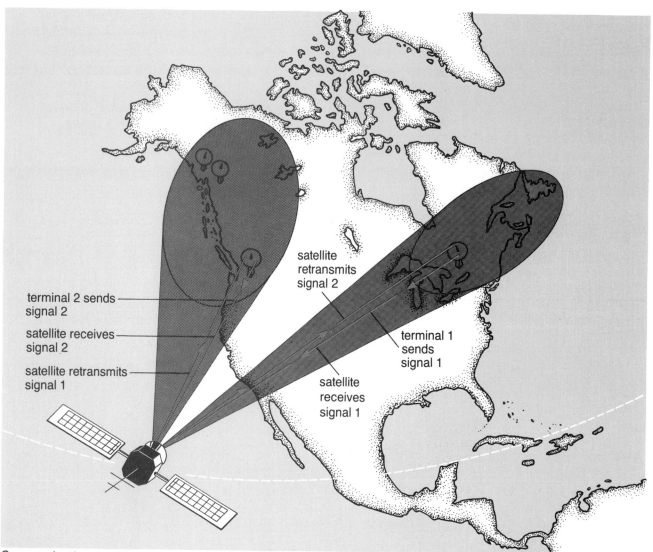

terminal 2 sends signal 2

satellite receives signal 2

satellite retransmits signal 1

satellite retransmits signal 2

terminal 1 sends signal 1

satellite receives signal 1

Communication satellites send and receive signals from different ground stations. The oval-shaped areas over which a satellite sends and receives signals are called its "footprints."

Did You Know?

The satellite *Alouette* was launched in 1962. It was designed and built in Canada to study a layer in the atmosphere that affects our communications. *Alouette* was designed to work for three years. It sent back information for a full 10 years before it was finally turned off.

The satellites receive signals from ground stations through their antennas. The satellites' antennas then redirect the signals back to other ground stations far away from the sending stations. In this way the signals do not have to travel close to the ground through bad weather conditions, or over mountains.

But communication satellites do much more than redirect radio and TV signals. One satellite, *Hermes*, helped bring educational programs to students who could not travel to attend classes. It also helped in medical emergencies. Physicians in southern Canada were able to send advice via the satellite on how to treat sick people in isolated regions where there were no physicians to help them.

Weather satellites carry cameras to take the pictures they send back to TV ground stations. These satellites follow an orbit that scans oceans and continents. Scientists can now study cloud formations, measure the speed of major weather systems, and predict more accurately what the weather will be like than they could before there were satellites.

Other satellites provide information on a whole array of different things: forest fires, icebergs, air pollution, crop conditions, and the migration of wild animals.

Probing

Research the *Anik* series of satellites. Every Canadian's day-to-day life is affected by them.

Working at Problem Solving

Sometime after 1995, a series of shuttles will head out to space to begin constructing Space Station Freedom. This international station will have a crew of eight from countries around the world, working together on some exciting projects. They will even grow their own vegetables in space!

Sherry Draisey is a structures engineer working at Spar Aerospace. Sherry was a member of the team of engineers who built the famous "Canadarm." She is now working on a new robotic arm that will help build Space Station Freedom.

Sherry and her co-workers are designing the new arm to be three times stronger than the Canadarm. "In space the arm will be able to pick up loads of 30 000 kg. On Earth it wouldn't even be able to pick up its own mass because of gravity."

Sherry says the new arm is quite a challenge to build. It's different from the Canadarm because it will have an extra joint to allow for more complex movements. "You have to test everything before it goes into space because once it's up there it's pretty hard to fix." Sherry tries out the arm on a device called a "flat floor," which has strain meters to test the strength and effects of different forces on the arm.

How did Sherry get involved in such an exciting project? "When I finished high school, I still didn't know what I was going to do. A teacher told the class that if we were looking for a challenge we should try engineering. I liked math and science, so I decided to take civil engineering courses at university."

Should someone be mechanically inclined to become an engineer? "Well, I'm not mechanical at all. I even have a hard time remembering which way to turn a screw driver." What Sherry enjoys is solving problems—and her job is full of them. Sherry uses her problem-solving abilities to help develop computer models of robotic arms. She and her co-workers can then analyse and improve the models before the final structure is built.

Checkpoint

1. Using what you have learned about the buoyant forces of liquids, predict which would be stronger, the buoyant force of water or the buoyant force of milk. Design an activity to test your prediction.
2. How are electrostatic forces
   (a) like magnetic forces?
   (b) like friction?
3. In Unit One you observed the movements of a fish.
   (a) List two forces that a fish would experience in its environment.
   (b) Give one example of how a fish overcomes the effect of each force.
4. Look again at the comic strip on page 180. Would it have been easier or more difficult for Andrew to knock Isaac Newton over if their masses had been equal? Explain your answer.
5. Using what you know about inertia, explain to a motorcycle passenger when it would be better to hold on to something in front like the driver, and when it would be better to hold on to something behind, like the bar across the back of the seat. Think about starting, stopping, and making turns. Be sure to give your reasons.
6. Using what you know about the inertia of objects at rest, explain why Thread A breaks when you pull slowly on the ring and Thread B breaks if you give the ring a sudden jerk.

7. Write the statements in Column A in your notebook and match with one or more examples from Column B.

Column A	Column B
Objects at rest have inertia.	When the bus driver slammed on his brakes, Jeff nearly fell flat on his face.
Objects in motion have inertia.	It took four of us to move that piano across the room.
For every action there is an equal and opposite reaction.	When Gina stepped off her skateboard it flew back and hit the wall.
Increase mass and you increase inertia.	I can push that chair across the floor but not when John is sitting in it.
	It was incredible! The magician pulled the tablecloth out from under all the dishes without breaking even one.

8. Choose one object in the classroom that is at rest. Describe the force (or forces) acting on the object. In which direction is the force(s) acting?
9. Satellites can stay in their orbits for years. Explain why this is possible.
10. Explain why a wall does not fall down when you push against it, using Newton's ideas about actions and reactions.

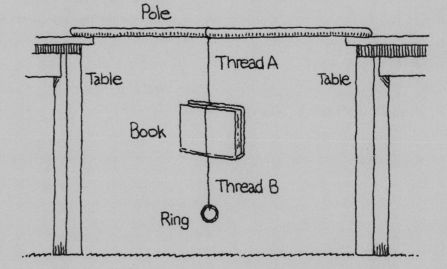

Focus

- All motion results from forces pushing or pulling on objects.
- Forces act in different directions and with different strengths.
- Gravity acts between all bodies that have mass.
- Mass is measured in kilograms (kg) and grams (g). Mass is measured with a balance scale. Mass remains the same wherever you are in the universe.
- Force is measured in newtons (N). Force is measured with a spring scale. Weight is a measure of the force of gravity. Weight can vary, depending on where you are in the universe.
- Friction occurs whenever two surfaces move against each other.
- Friction causes wear, slows down or stops motion, and generates heat.
- An electrostatic force can result when two surfaces move against each other.
- Water and air exert an upward buoyant force.
- An object at rest will stay at rest unless an unbalanced force acts on it. Then the object will move in the direction of the force. An object in motion will not change its speed or direction unless an unbalanced force acts on it. This tendency of objects to stay at rest or continue in motion is called inertia.
- Inertia increases if mass is increased.
- For every action there is an equal and opposite reaction.
- The combined effect of a force causing forward motion and the force of gravity is a curved path. Under certain conditions the effect may cause a satellite to orbit.
- Because our country is so large, we need ways to communicate over long distances. Canadian scientists have developed satellites that provide us with excellent methods of communication.

Backtrack

1. (a) List as many different kinds of forces as you can.
   (b) Which of these forces can start objects moving without touching them?
2. List three factors that affect the amount of friction between two surfaces when at least one of the surfaces is moving.
3. Think of some of the items you tested with your force meter. Suggest ways to lessen friction and lessen the force needed to operate these items.
4. Which of these statements show a lack of understanding about the difference between mass and weight? Explain what is wrong with the statement in each case.
   (a) Your mass on the moon is 1/6 of what it is on Earth.
   (b) "My weight was 2 kg more after the summer holidays."
   (c) "According to my spring scale, the mass of your parcel is 400 g."
   (d) The astronauts each had a mass of about 700 N on Earth.

(e) "I am going to take a balance scale with me to Venus to see if this rock weighs the same there as it does on Earth."

5. (a) Describe how to give an object an electrostatic charge.
(b) Which conditions make it easier to "charge" an object?

6. (a) Describe how to magnetize an object.
(b) List as many materials as you can that can be magnetized.

7. Explain why a satellite stays in orbit.

8. You walk into the classroom and see this set-up. What do you infer is behind the cloth?

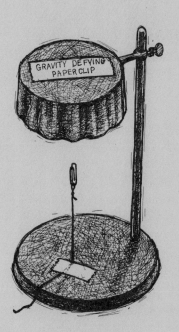

Synthesizer

9. Design a vehicle that can be made to move by the action-reaction of an inflated balloon or a tightened elastic band.

10. A salt and pepper mixture is first placed on paper as in the illustration. A plastic ruler is rubbed with wool or flannel cloth, and is then held over the mixture.
(a) Predict what will happen to the pepper and the salt.
(b) Explain what force would cause this effect.

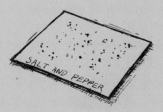

11. Why do sprinters always start a race in a crouching position?

12. Think about a bicycle.
(a) What has been done to the part of the handlebars that you grip? Explain why.
(b) Describe the surface of the pedals. Explain why pedals have this surface.

13. (a) How is the discus thrower's arm like the force of gravity?
(b) When the discus thrower lets go of the discus, in what direction will it go?

14. You have probably seen and heard of dozens of people who suffered a bad fall because they "slipped."

Maybe you are one of them! This is an opportunity to take a close look at possible accident sites all around you—at school, at home, in the community, on a farm, and so on. These "slips" happen on stairs, sidewalks, hard floors, and in the bathtub. Often they happen when surfaces are wet or slippery. Make a list of places at school or at home that could be dangerous. For each, describe a method of using friction to make the place less dangerous. Do you have other ideas as well to make these places safer? One of your ideas might save someone from a terrible fall.

15. Electrostatic air cleaners remove dust particles from the air. Design an electrostatic air cleaner. Make a drawing of it and explain how it would work.

Temperature and Heat

Tracy has been training for this race all year. She has practised in all kinds of weather, on all kinds of surfaces, by herself and with others. Now, within sight of the finish line in the final race, she has collapsed. What has gone wrong? Why has her body given out so suddenly?

Unfortunately for Tracy, today is hot and sunny. The heat of the day and the heat produced by her running have made her body temperature rise above normal. Her brain cells have responded to this overheating. Their response? To make her become unconscious. Her trainer must work fast to cool her body down.

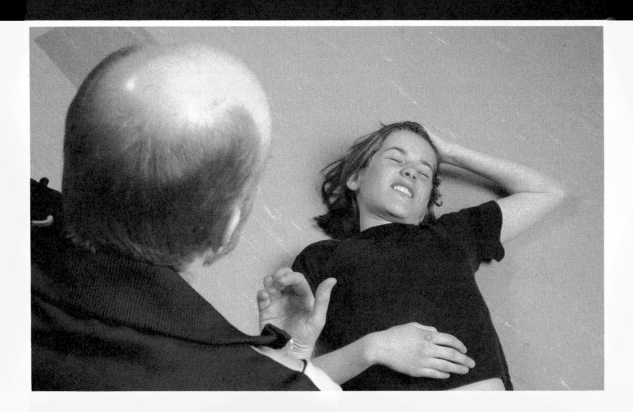

As you learned in Unit One, correct body temperature is essential for animals to carry out their normal functions. Does temperature also affect non-living things? How can you measure temperature? What do "hot" and "cold" mean? What is the difference between temperature and heat? Can heat be measured? How is heat produced? These are some of the questions you will investigate in this Unit.

Hot and Cold

We all know when things are hot or cold. Or do we? And do we know *how* hot or *how* cold?

You use objects made of glass every day. A glass jar taken from the refrigerator is cold. A glass pie plate taken from the oven when the pie is cooked is hot. So is the glass in a light bulb that has been turned on for a few minutes. But the glass in a light bulb stored in a cupboard is neither hot nor cold.

You know from experience when things are too hot to touch. But when you are handling objects that *are* cool enough to touch, can you rely on your senses to tell you how hot or cold the objects are? Just how reliable is your sense of touch? This Topic begins with an Activity that tests whether you can judge how hot or cold an object is by touch.

You could use your sense of sight to infer that the glass in this photograph is hot. When glass gets very hot its appearance changes and it begins to flow. Because the glass in the photograph is far too hot to touch, the glass worker is using a tool to shape the glass. The glass worker has learned to tell by looking at the glass when it is hot enough to be shaped.

Hot or Cold?

1. Try this for yourself. Put one hand in a beaker of hot water (not too hot!) and one hand in a beaker of cold water. Hold them there for at least one minute. Then put both hands in a beaker of lukewarm water. What messages does your brain receive from each hand?

2. When your hands have returned to normal temperature, find out how well you can judge whether objects are hot or cold. Touch various objects in your classroom. Try your desk, a pen, a wooden ruler, a metal ruler or doorknob, a book, an eraser, or any other convenient object in the classroom.

3. In your notebook record whether each object feels warmer, the same, or colder than your hand.

Which objects feel cold? Are they really colder than the other objects you have touched?

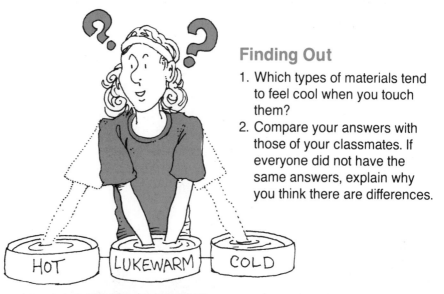

Finding Out

1. Which types of materials tend to feel cool when you touch them?

2. Compare your answers with those of your classmates. If everyone did not have the same answers, explain why you think there are differences.

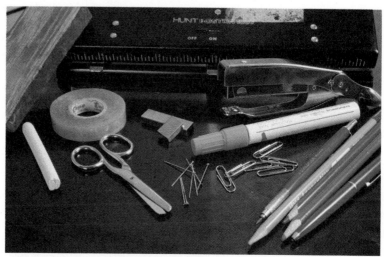

Estimating Temperature

As you can see, it's difficult to judge exactly how hot or cold something is just by touching it. A better way is to find out what the temperature of the object is. For now, you can think of temperature as the measurement of how hot or cold something is. We usually measure temperature in degrees Celsius (°C). For example, the temperature of the palm of your hand is probably about 35°C. In the next Activity you will discover how much you know about temperature in a number of different situations.

Estimating Temperatures

Problem

How accurately can you estimate temperatures?

Materials

28 index cards

Procedure

1. In a group, copy the items listed in Table 1 onto the index cards (one item per card). Work first with cards 1 to 14. Set aside the temperature cards (15 to 28) until Step 4.

Table 1

1. Boiling temperature of water
2. Body temperature of healthy human being
3. Hot tea
4. Average room temperature
5. Ice cream
6. Coldest temperature possible
7. Comfortable bath water
8. Temperature of air in oven when roast is cooking
9. Coldest air temperature recorded on Earth
10. Surface of the Sun
11. Hottest air temperature recorded on Earth
12. Freezing temperature of water
13. Air in a refrigerator
14. Interior of the Sun
15. −273°C
16. −88°C
17. 80°C
18. −10°C
19. 100°C
20. 20°C
21. 58°C
22. 15 000 000°C
23. 37°C
24. 160°C
25. 0°C
26. 40°C
27. 6000°C
28. 7°C

2. Discuss each of these cards with your group and estimate the temperature of the item on each card. Now sort the cards into three piles: "low temperatures," "average everyday temperatures," and "high temperatures."
3. Now discuss each pile of cards separately. Arrange the cards in each pile in order, from coldest to hottest estimated temperature.
4. Next take the temperature cards and match them with cards 1 to 14. Change the order of cards 1 to 14 until you have made your best estimates about the temperature in each situation.
5. In your notebook, record the temperature that your group has decided upon for each of the cards 1 to 14 from the coldest to the hottest.

Finding Out

1. Your teacher will supply the actual temperature for cards 1 to 14. On the back of each card, write the actual temperature, then compare it with the temperature your group estimated.

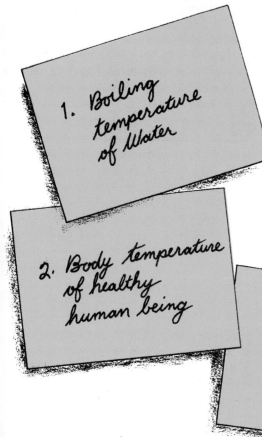

1. Boiling temperature of Water

2. Body temperature of healthy human being

3. Hot tea

A LIQUID CAN BE POURED. SALT CAN BE POURED. IT FLOWS. SO I SUPPOSE SALT IS A **LIQUID**.

A LIQUID TAKES ON THE SAME SHAPE AS ITS CONTAINER. IF I POUR WATER INTO A CUP, THE WATER TAKES THE SHAPE OF THE CUP.

NOW, IF I FILL A CUP WITH SALT, I CAN GET A BIG PILE OF SALT ABOVE THE TOP OF THE CUP. THE SALT DOES **NOT** TAKE ON THE SHAPE OF THE CUP. SO SALT IS NOT A LIQUID

States of Matter

In Activity 4-2 you estimated the temperature of air in an oven, and water in a bath. You know that air is different from water, and water is different from steel. Air is a **gas**, water is a **liquid**, and steel is a **solid**. All substances can be grouped into one of these three **states of matter**—gas, liquid, or solid.

To understand how a gas, a liquid, and a solid differ, it helps to know the meaning of volume. The **volume** of a material or an object is the amount of space it occupies. For example, think of all the space the vehicles in a parking lot take up. Now think about a bus and a car: the bus has a larger volume than the car.

A gas has the same volume and shape as the container it is in; a gas can flow. A liquid has a set volume, but it has the same shape as the container it is in; a liquid can also flow. A solid has a set volume and shape; it cannot flow as a gas or a liquid do.

The characteristics of gases, liquids, and solids. There is the same amount of gas in each of the two containers in (a). There is the same amount of liquid in each of the two containers in (b), and the same amount of a solid in the two containers in (c).

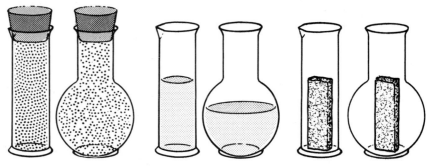

Heating and Cooling Gases

Air is a mixture of gases. The main gases that make up the air you breathe are oxygen, nitrogen, and carbon dioxide. Other gases you may have heard of are hydrogen, helium, and neon. Since air is easy to use in class, you will be using it in the next Activity to test what happens to a gas when its temperature changes.

Activity 4-3

Heating a Pop Bottle

Problem

What happens to the air in an empty pop bottle as it is heated?

Materials

empty pop bottle (small size)
25-cent piece
stop watch or other timer
a few drops of water

Finding Out

1. What causes the popping?
2. Draw a bar graph of your results.
3. Does the popping increase, decrease, or stay the same as time goes on?
4. Compare your results with the results of other groups.
5. Explain why the rate of popping changes.

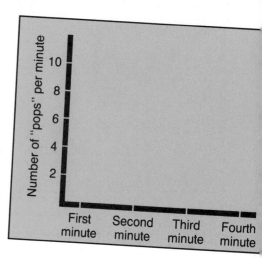

Procedure

1. Decide who will use the equipment and who will do the timing and recording.

2. Moisten the mouth of the pop bottle and one side of the coin. Place the coin wet-side down on the mouth of the pop bottle.

3. Wrap your hands around the bottle and count the number of "pops" in the first minute, second minute, third minute, and fourth minute.

Expansion and Contraction

When air is heated, it spreads out and occupies more space; in other words, its volume increases. This increase in volume is called **expansion**. You can see the effect of expansion when a hot-air balloon is heated.

When air is cooled, the reverse happens: the air occupies less space, and its volume decreases. This decrease in volume is called **contraction**.

(a) A heater is being used to expand the air in a hot-air balloon.

Heating and Cooling Liquids

Gases expand and contract. Do liquids also expand and contract? Find the answer to this question by doing the next Activity.

Activity 4-4

Heating Water

Problem

How does heat affect the volume of a liquid?

> **CAUTION:** Do not attempt to insert or remove glass tubing from the stopper. Be careful when using a source of heat.

Materials

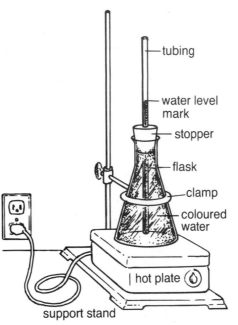

tubing

water level mark

stopper

flask

clamp

coloured water

hot plate

support stand

The hollow glass tubing passes through the hole in the stopper.

Procedure

1. Set up your equipment as shown in the illustration.
2. Mark the water level in the glass tubing with a piece of tape.
3. Heat the water in the flask. (Do not allow the water to get hot enough to boil.) Mark the water level again.
4. Allow the flask to cool and observe any change in the water level.

Finding Out

1. Did the water expand or contract when it was heated? How do you know?
2. Was this change permanent?
3. Do liquids behave like gases when they are heated or cooled? Explain.

Finding Out More

4. Predict what would happen to the water level in the tubing if you placed the flask in the refrigerator.

(b) The balloon, inflated with hot air, then rises through the cooler air around it. When the balloon is ready to descend, the heater is turned off and the air in the balloon cools down. What happens to the air in the balloon as it cools?

Heating and Cooling Solids

Ball and ring both at room temperature.

Ball is heated in a hot flame.

One definition of science is "the search for patterns in nature." Have you seen a pattern in the effect of heat on liquids and gases? Is this pattern true for solids as well?

What happens to metals when they are heated? A ball and ring apparatus will show you the effect of heat on a metal.

At room temperature the ball will pass through the ring. Predict what would happen if you
(a) heated the ball;
(b) heated both the ball and the ring;
(c) cooled the ring under cold water when the ball was at room temperature.

If a metal measuring tape is used to check distance, would you prefer to try setting a record on a hot or cold day?

Why do railway tracks have gaps between the sections?

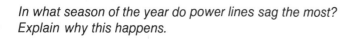

In what season of the year do power lines sag the most? Explain why this happens.

Comparing Different Materials

You have seen that materials expand when heated and contract when cooled. Do all materials expand by the same amount when they are heated? Examine this question, first using liquids, and then using solids.

Activity 4-5

Comparing Different Liquids: Demonstration

Problem

Do all liquids expand by the same amount when they are heated?

Materials

3 different liquids (water, vegetable oil, and burner fuel or ethyl alcohol), all at room temperature
hot plate
safety goggles
3 large test tubes
3 one-holed rubber stoppers, each with a long piece of glass tubing inserted
500-mL beaker
stirring rod
support stand and clamps
masking tape

Procedure

CAUTION: Exercise care when using a hot plate or other source of heat. There must be no open flames during this Activity!

1. Set up the equipment as shown in the illustration. The liquids in the test tubes should all be at room temperature and the water in the beaker should be taken from the cold water tap. Adjust the liquids so that they all rise to about the same height in the glass tubing, just above the stoppers. Mark these levels—the starting positions—with pieces of tape.

support stand

glass rods in which three different liquids will rise

stirring rod

masking tape to hold test tubes together

water bath

hot plate

2. Turn on the source of heat. Observe the liquids in the tubing. Carefully stir the water in the beaker.

CAUTION: Do not let the water in the beaker boil. If one liquid comes near the top of the tube, turn off the heat source.

3. Record what happens to the level of each liquid.

Finding Out

1. What has happened to the volume of each liquid?
2. What is your answer to the problem: Do all liquids expand by the same amount when they are heated?

Finding Out More

3. Explain why all the liquids should be at room temperature at the beginning of the Activity.

Heating and Cooling Different Solids

Now consider the effect of heating and cooling on different solids. Try this thought activity. Think about the following situation: Imagine you have a glass jar with a metal lid. The lid is too tight to get off.

How could you use heat to get the lid off? The method shown in the illustration often works because the metal expands more than the glass when they are both heated by the same amount.

From this thought activity you might infer that different solids expand by different amounts when heated. Is this always the case? Try another thought activity, this time about teeth.

Suppose that you had several fillings in your teeth, and that these fillings expanded at a faster rate when heated than the teeth themselves. What might happen when you had a drink of hot chocolate? Now suppose that the fillings also contracted faster than the teeth themselves when cooled. What might happen when you bit into some ice cream? What can you infer from this thought activity about the materials dentists use for filling teeth?

You need more evidence before you can make predictions about the different rates of expansion and contraction of solids. You will have more opportunities to observe the expansion and contraction of solids in the next two Topics.

This X-ray photograph of all the teeth in a patient's mouth was taken by a machine that moves slowly around the front of the patient's head. The X-rays pass easily through flesh; in these areas the film is dark. They pass less easily through bones and teeth, so those areas look lighter. The X-rays cannot pass through dental fillings. The fillings appear as white areas. To prevent the film being blurred, the patient must remain absolutely still while the photograph is taken. To help keep the mouth still, the patient holds a piece of plastic (seen faintly in the photograph) between the front teeth.

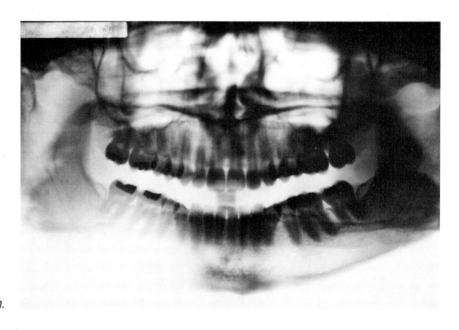

Measuring Temperature

Temperature is important to you in many ways. Have you ever had soggy French fries? If the oil for cooking the fries isn't hot enough, the fries will be soggy. When you bake a cake, the oven temperature must be just right or the cake will be flat and hard. A car is designed so that a warning light comes on if the engine gets too hot. In the steel-making industry, the temperature in the furnace must be high enough to melt the iron and other substances used to make the steel alloy. You can tell when it is hot in summer, or cold in winter. But if you need to know *how* hot or *how* cold, you find out from the weather report.

Galileo was famous for many scientific achievements. Besides making thermometers, he invented the telescope and used it to observe things people had never seen before, such as the Moon's mountains and valleys. He was also very interested in the ideas of Copernicus, a Polish scientist, who had suggested that the Earth revolves around the Sun. Before this time, most people thought that the Sun revolves around the Earth, because the Sun appears to move across the sky.

Several hundred years ago, people had no way of measuring the exact temperature—for cooking, or for anything else. They had ways of estimating temperatures. But did "cold" on a sunny island in the south mean the same as "cold" in the north of Canada in winter? An exact way of measuring temperature was needed.

An Italian scientist, Galileo Galilei, who lived in the 1600s, invented the first device for measuring temperature. He called this device a thermometer, from *thermo* meaning heat, and *meter*, meaning measuring device. You have seen that solids, liquids, and gases expand when heated and contract when cooled. Galileo used his knowledge of expansion and contraction in his thermometer. His first thermometer contained mostly air. Can you see what happens in this air thermometer when the temperature rises?

Galileo Galilei and the air thermometer he invented. There is a bubble of liquid in the thermometer. When the temperature rises, the air in the thermometer expands. The air pushes against the bubble, causing it to rise. The level of the bubble can thus be used to give a temperature reading.

Building Your Own Thermometer

In the next few Activities, you will make thermometers yourself. Before you do, you should learn all you can about the type of thermometer used in science classrooms. In Galileo's thermometer air expands and contracts, but in the classroom thermometer it is a liquid that expands and contracts.

The Laboratory Thermometer

Problem

What is a laboratory thermometer like and how does it work?

Materials

laboratory thermometer (with scale from below 0°C to above 100°C)

> **CAUTION:** A glass thermometer breaks easily. Handle it carefully.

Procedure

1. Examine the thermometer carefully. Notice where the glass is thick and where it is thin. Draw a full-size diagram of the thermometer in the centre of a page in your notebook.
2. Put these labels on your diagram:
   - liquid (coloured alcohol)
   - bore (the fine opening through which the liquid moves)
   - thick glass
   - bulb (storage space for the liquid)
   - thin glass
   - the scale
   - freezing point of water (0°C)
   - boiling point of water (100°C)
   - lowest temperature (state what it is)
   - highest temperature (state what it is)
   - average room temperature (20°C)
   - average normal body temperature (37°C)

3. (a) What happens to the temperature reading if you place your hand around the middle of the thermometer?
   (b) What happens to the temperature reading if you put your hand around the bulb?

Finding Out

1. The bulb is an important part of a liquid thermometer. Give reasons why.
2. Which part of the thermometer has the thinnest glass? Why do you think it is made like this?

Finding Out More

3. (a) How many Celsius degrees are there between the freezing point of water and the boiling point of water?
   (b) Celsius degrees are also called centigrade degrees by some people. Explain why. (Hint: What does *centi* mean?)
4. Why should you not put this thermometer into a liquid at 300°C?

Extension

5. If other types of liquid thermometers are available, compare them to the one you used in this investigation.
   (a) How are they different?
   (b) How are they similar?
6. Find out why mercury is no longer used in making thermometers for use in school.

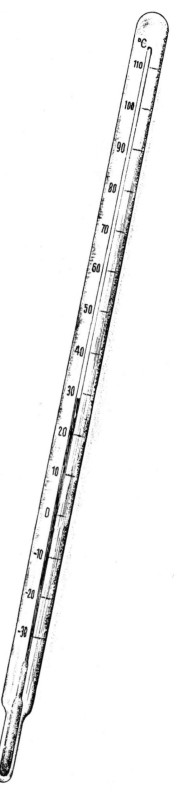

The temperature scale you are most familiar with is the Celsius scale. The Celsius scale was invented by Anders Celsius in Sweden in the 1700s. To make his scale he marked on his thermometer the freezing point of pure water as 0 and the boiling point of pure water as 100. Then he divided the space between these points on the thermometer evenly into 100 "degrees" (symbol °). Temperatures below 0°C and above 100°C can be marked on the scale.

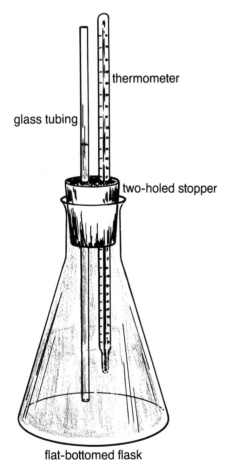

glass tubing

thermometer

two-holed stopper

flat-bottomed flask

Activity 4-7

Making a Liquid Thermometer

Problem

How can you make a liquid thermometer to measure the temperature in a room?

Materials

two-holed stopper with glass or plastic tubing already inserted through one hole, and a laboratory thermometer already inserted through the other hole
coloured water (at room temperature)
flat-bottomed flask
masking tape
sink with hot and cold running water

CAUTION: Do not try to remove the thermometer or the tubing from the stopper.

Procedure

1. Completely fill the flask with coloured water at room temperature. Put the stopper with the tubing and the thermometer into the top of the flask. You want to have the liquid rise about half-way up the tubing. You may have to add or remove some of the coloured water to get it at the right level.
2. Place the flask in the sink so that you can run water over it. You can lower the temperature in the flask by running cold water over it; when you want a

higher temperature, you can run hot water over the flask.
3. Use the laboratory thermometer to calibrate your new thermometer—the tubing. (Do you remember how to calibrate from Unit Three?) Use masking tape to mark the level of the water at certain temperatures. For example, try 15°C, 20°C, and 25°C or perhaps 16°C, 18°C, 20°C, 22°C, and 24°C.
4. After you have finished calibrating the thermometer, *do not touch the rubber stopper.* Can you explain why?
5. Move the thermometer very carefully to a place in the classroom where you can leave it set up for several days. Check it each day to see if it is showing the same temperature as the laboratory thermometer.

Finding Out

1. Compare the thermometer you have made with the laboratory thermometer.
   (a) In what ways are they similar?
   (b) In what ways are they different?
2. What problems did you notice when you were
   (a) making the liquid thermometer?
   (b) calibrating the thermometer?
   (c) using the thermometer?
3. Suggest ways to improve the design of your new thermometer.

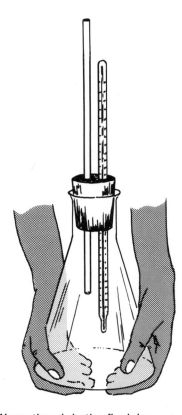

Warm the air in the flask by holding the flask upright in your hands for several minutes.

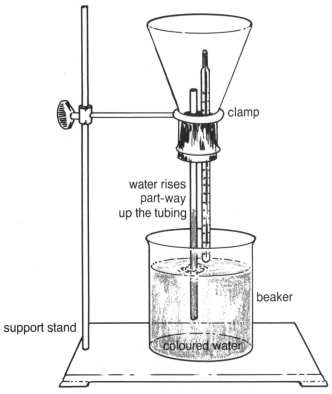

clamp

water rises part-way up the tubing

beaker

support stand

coloured water

Turn the apparatus upside down so that the tubing extends half-way into the coloured water.

Extension

4. Make a gas thermometer using the materials and procedure shown in the illustrations.
- It may take more than one try to get the liquid to rise about half-way up the tubing at room temperature.
- Calibrate the gas thermometer as you did the liquid thermometer.
- Over several days find out how well the liquid thermometer and the gas thermometer work.
- Make a copy of Table 2 in your notebook and list the advantages and disadvantages of each thermometer.

Table 2

	ADVANTAGES	DISADVANTAGES
Liquid thermometer		
Gas thermometer		

Using Solids to Measure Temperature

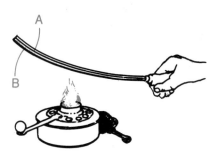

Which metal, A or B, expands and contracts more when the strip is heated and cooled? Explain how you know.

In Topic One you thought about a glass jar with a metal lid—an example of different solids that expanded different amounts when heated. You can see another example of two different solids expanding different amounts in a device called a **bimetallic strip**. In this device two strips of different metals are bonded together.

A bimetallic strip acts in an interesting way when heated or cooled. Look at the bimetallic strip in the illustration. When it is heated, the strip bends. This happens because one of the two metals in the strip expands more than the other. You can use this effect to make a "solid" thermometer.

Activity 4-8

A Solid Thermometer: Demonstration

Problem
How can a bimetallic strip be used to measure temperature?

Materials
bimetallic strip attached to a handle
heat source with a flame
scale drawn on paper or cardboard as in the diagram
thermometer (optional)

> **CAUTION:** Be careful when using an open flame.

Procedure
1. Examine the bimetallic strip and describe how it is made.
2. Hold the bimetallic strip in the flame for a few minutes. Record what you observe.
3. Hold the strip under cold running water for several minutes. Record what you observe.
4. Explain how you could hold the bimetallic strip so that it can be used with a scale. Try it to see if it works.

Finding Out
1. Compare the solid thermometer with a laboratory thermometer that uses liquid.
   (a) How are they similar?
   (b) How are they different?

Finding Out More
2. If possible, state which two metals make up the bimetallic strip. Which metal expands more when the strip is heated? How do you know?
3. Is there any evidence to support the statement: "If metal A expands more than metal B when heated, then metal A also contracts more than metal B when cooled"?

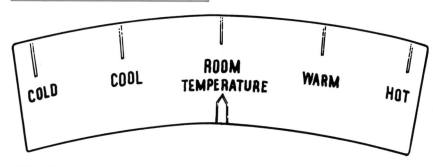

1. Marco slowly enters a swimming pool. He thinks the water is too cool for swimming. After a couple of minutes in the pool, he thinks the temperature is just fine.
   (a) Explain why he changed his mind.
   (b) How does this situation compare with the situation you encountered in Activity 4-1?

2. Tamara gets out of bed in the morning and walks across a rug onto a tile floor in the bathroom. The tile feels much colder than the rug, but Tamara infers they are at the same temperature. Do you think she is right? Explain.

3. A magician at the fall fair had five copper pennies, each with a different year marked on it. She put the pennies into a hat and asked someone in the audience to pick out one coin. Then she said, "Pass the coin around so everyone can see what year is on it. Then put the coin back into the hat and I'll try to pick out the same coin." When she reached into the hat, there was a brief pause and then she pulled out the correct penny. How did she do it? (You might want to try this yourself.)

4. (a) What are the three states of matter?
   (b) Give two examples of substances in each state of matter.

5. In Activity 4-7 you used water to make a thermometer. Would water be an efficient liquid to use in a thermometer for general use in homes and schools? Why or why not?

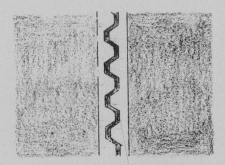 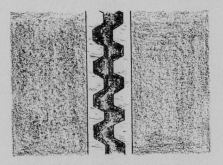

6. Look at the two diagrams above of the paved surface of a highway overpass. Which diagram represents the paved surface in the summer? Explain your answer.

7. A table-tennis ball was partly crushed by accident. To restore it to its original shape, Marim decided to apply what she had studied in science class. She put the ball into boiling water for one minute. Why did she think of doing this? Do you think her method worked? Explain your answer.

8. Assume you are a technologist responsible for making bimetallic strips. You discover that aluminum expands more than brass, and brass expands more than steel. You make three bimetallic strips, as shown below.

(a) Which way will each strip bend when heated?
(b) One strip will bend much more than the others. Which one? How do you know?

9. Ordinary glass breaks easily when it is heated or cooled quickly. Most beakers and flasks in science laboratories are made of Pyrex glass. This type of glass can be heated or cooled quite quickly without breaking. Do you think Pyrex glass expands more or less than ordinary glass when it is heated? Explain why you think so.

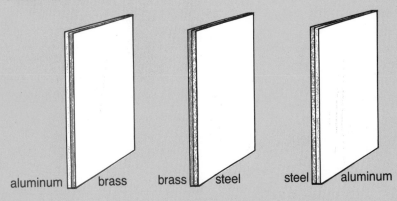

aluminum   brass   brass   steel   steel   aluminum

Special Purpose Devices

Why do people need to measure temperature? Think about all the ways temperature measurement is important in your home. Can you also think of ways it might be important in industry? in health care?

Solids, liquids, and gases all expand when heated and contract when cooled, and any one of these might be used in a thermometer. The choice depends on the application or the way the thermometer will be used. In this Topic you will investigate some special types of thermometers. You'll also find out about some of the ways scientists have developed for measuring the temperature of objects that are out of reach, such as the Sun.

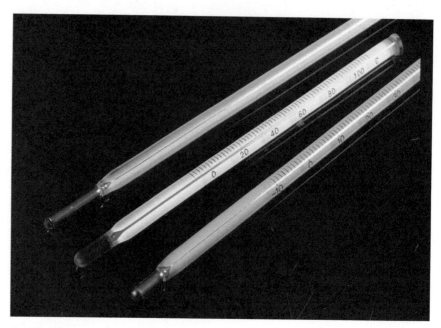

Some laboratory thermometers.

Different Types of Thermometers

Problem

What are the differences among various types of thermometers?

Materials

several different thermometers (gas, liquid, and/or solid)

Procedure

1. Examine and sketch each thermometer. Label your sketches.
2. Give each thermometer a number. Make a table similar to Table 3 in your notebook.

Refer to your sketches and, if necessary, go back and observe each thermometer carefully in order to fill in the columns in the table.

Finding Out

1. For each thermometer, explain why its structure, design, and materials make it suitable for the function it is used for.
2. For each thermometer, explain why a solid, a liquid, or a gas was chosen to make the thermometer.

Table 3

CHARACTERISTICS OF THERMOMETER	THERMOMETER 1	THERMOMETER 2
Length (cm)		
Width (cm)		
Lowest temperature (°C)		
Highest temperature (°C)		
Substance(s) in the thermometer that expand and contract		
Thermometer's use		
Advantages		
Disadvantages		

Thermometers for Special Uses

The specialized thermometers shown below can record a temperature; the reading of the temperature can be taken long after the temperature was recorded. If you can obtain thermometers like these, try them in hot and cold water. Who do you think might need to use maximum and minimum thermometers?

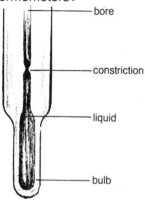

The **clinical thermometer** shown here measures body temperature. It must be sensitive to very small changes in temperature, so it is made to measure temperatures within a few degrees of normal body temperature (37°C). Note the very narrow part of the bore just above the bulb, called the constriction. As the liquid is warmed by the patient's body heat, it expands past the constriction to show the temperature reading. After the temperature has been taken, the liquid cools and contracts. But the liquid cannot move back past the constriction; it must be forced back by shaking the thermometer downwards. Can you think of the reason for having a constriction in a clinical thermometer?

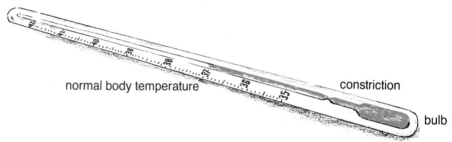

normal body temperature · · · · · · · · · · constriction · · · · · bulb

The maximum thermometer works like a large clinical thermometer. Explain why this thermometer has a constriction in its bore.

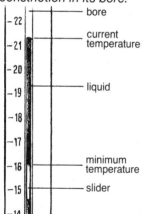

In a minimum thermometer, as the temperature drops and the liquid contracts, it pulls a solid "slider" down. If the temperature goes back up, the liquid moves past the slider, which remains in place. To reset the slider, the thermometer should be held upright.

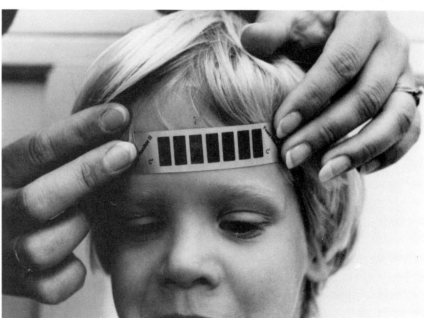

Body temperature is sometimes measured using a liquid crystal thermometer. The flexible strip applied to the forehead contains a liquid crystal colour chart that indicates temperature within about 10 s. This type of thermometer is cheaper and easier to use than a clinical thermometer, but it is not as accurate.

A **thermostat** measures temperature in an appliance or in a room. It switches the appliance on or off at a pre-set temperature. For example, a thermostat can be set to control a furnace to keep room temperature at 20°C. When the temperature in the room rises above 20°C, the thermostat causes the furnace to turn off. When the temperature drops below 20°C, the thermostat causes the furnace to turn back on.

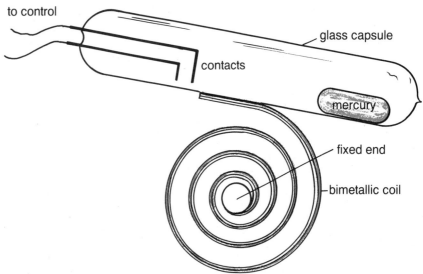

One type of furnace thermometer uses a bimetallic strip and a mercury switch. When the bimetallic strip coils up, the glass capsule tilts downward. The mercury slides to the left and touches the two contacts, and the furnace turns on.

Probing

Thermostats containing bimetallic strips are used in electric irons.

1. What other appliances can you think of that are controlled by thermostats?
2. How can thermostats be used to conserve energy?
3. If there is a furnace thermostat in your home, and you have the permission and help of an adult, carefully remove the cover of the furnace thermostat. Try to figure out how it works. Draw a sketch showing the main parts, and describe as much as you can about its operation.

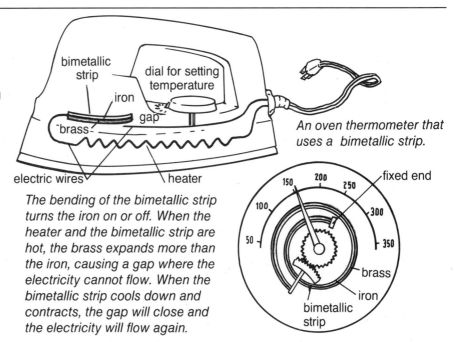

The bending of the bimetallic strip turns the iron on or off. When the heater and the bimetallic strip are hot, the brass expands more than the iron, causing a gap where the electricity cannot flow. When the bimetallic strip cools down and contracts, the gap will close and the electricity will flow again.

An oven thermometer that uses a bimetallic strip.

A **thermocouple** is a thermometer that uses electricity to measure temperature. A thermocouple contains two different metal wires. The wires are joined (coupled) at one end. The other ends are joined to a meter that measures electricity. When two different metals touch, a tiny amount of electricity is produced. The amount of the electricity varies with the temperature. By measuring the amount of the electricity that flows through the meter, you can discover the temperature of the metals.

Thermocouples are used to measure temperatures in places where people cannot go to read a thermometer, such as inside a smokestack. Another feature of thermocouples is that they can measure higher and lower temperatures than liquid thermometers can. And thermocouples can be connected to computers to record temperatures.

Resistance thermometers also use electricity to measure temperature. They can be used at even higher or lower temperatures than thermocouples can. In a resistance thermometer, electricity flows through a platinum wire. Platinum is a metal that allows electricity to flow very easily. As with a thermocouple, the amount of electricity depends on the temperature. By measuring the amount of electricity that flows through the wire, you can determine the temperature.

You've probably noticed how a heating element on an electric stove goes from grey to a glowing red as it gets hotter. Hot objects give off light as well as heat. The brightness of the light and the kind of light changes with the temperature of the object. An **optical pyrometer** analyses these changes and uses the results to measure temperature. Unit Three explained how space vehicles become extremely hot when they return to Earth—so hot that ordinary metals would melt at these temperatures. Everyday thermometers and even thermocouples and resistance thermometers are useless for measuring temperatures as high as these, so optical pyrometers are used.

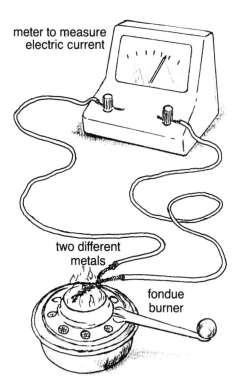

A thermocouple can be attached to an electric meter to measure temperature. Here, the thermocouple is measuring the temperature of a flame.

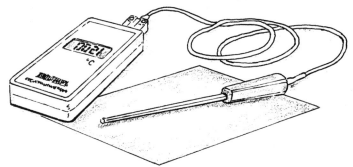

A platinum resistance thermometer can measure temperatures from about -200°C to about 1200°C.

Assume you have a camera with infrared film in it. Describe how you would use the film to answer these questions.

1. What parts of the human body allow the most heat to escape in cold weather? How can you use this information to dress properly when hiking on a cold winter's day?
2. What parts of a home allow the most heat to escape in winter? How can you use this information to improve the insulation in the home?

Thermographs measure temperature in the ordinary, everyday range. A thermograph is a special type of photograph. Like an optical pyrometer, it depends on light to measure temperature, but the light it uses is **infrared light**, which is invisible to human eyes. A thermograph shows very small differences in temperature—differences that cannot be measured with other types of thermometers.

A thermograph of someone holding a hot cup.

The thermograph shows the same house as the photograph. The blue colour indicates the areas of greatest heat loss. What other areas also show that heat is being lost?

What Is Heat?

You would probably predict that if you add heat to something, its temperature would go up. But there are times when adding heat may not increase the temperature! Why? What is heat? What makes the difference between cold water and hot water? These are some of the questions you will explore in this Topic.

The first problem to solve is: What is the difference between temperature and heat? We can say the temperature of an object is hot, cold, or warm, or we can be more exact and state its temperature in degrees Celsius. Either way, we are describing the conditions of one object at one particular time when we refer to its temperature.

When we talk about heat, however, we should always think about *two* objects. Heat always transfers *from* one object *to* another object. For example, if you put your hand in hot water, some heat transfers from the water to your hand. What is being transferred is *heat energy*. In the next Activity you will investigate this transfer of energy.

Workers must wear special clothing to protect themselves from the heat.

Mixing Hot and Cold Water

Problem

When hot and cold water are mixed,
(a) is energy transferred from the hot to the cold water or from the cold to the hot water?
(b) can the temperature be predicted?

Materials

2 large plastic cups or beakers
thermometer
100-mL graduated cylinder
cold water
hot water from tap

Procedure

1. (a) Measure 50 mL of hot water and record its temperature. Then measure 50 mL of cold water and record its temperature.

50 mL hot water

50 mL cold water

CAUTION: Do not stir while the thermometer is in the water.

(b) Add the hot water to the cold water, stirring the mixture with a stirring rod. Record the temperature of the water.

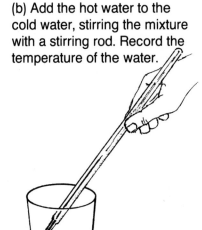

100 mL water

2. (a) At the same temperatures as in step 1 (a), predict what the temperature will be when you add 100 mL hot water to 50 mL cold water.
(b) Try it and record the temperature.

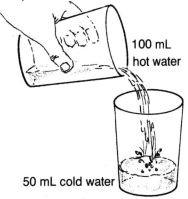

100 mL hot water

50 mL cold water

3. (a) Predict what the temperature will be when you add 150 mL hot water to 50 mL cold water, once again at the same temperatures as in step 1 (a).

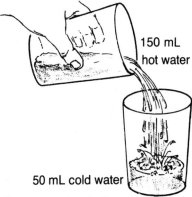

150 mL hot water

50 mL cold water

4. Draw a table similar to Table 4 to record your results.

Finding Out

1. Temperature is a measure of energy level. Which has a higher energy level, hot or cold water?
2. When hot and cold water are mixed, is the energy transferred from the hot water or from the cold water? How do you know?
3. Which has more energy to give off, a small amount of hot water or a large amount of hot water? How do you know from this Activity?

Table 4

STEP 1	TEMP. °C	STEP 2	TEMP. °C	STEP 3	TEMP. °C
50 mL hot water 50 mL cold water Mixture		100 mL hot water 50 mL cold water Mixture		150 mL hot water 50 mL cold water Mixture	

Finding Out More

4. Keeping the same temperatures of hot and cold water as you used before, what do you think the final temperature would have been if you had added 200 mL hot water to 50 mL cold water?

5. Keeping the same temperatures of hot and cold water, what do you think the final temperature would have been if you had added 100 mL hot water to 100 mL cold water?

6. Would your results have been different if you had poured the cold water into the hot water? Try it and see.

7. Predict what the temperatures of the following mixtures would be, using the same starting temperatures of hot and cold water as you used before:
   (a) 200 mL hot water and 50 mL cold water
   (b) 50 mL hot water and 100 mL cold water
   (c) 50 mL hot water and 150 mL cold water
   List your predictions, prepare a table for your data, and test your predictions.

Temperature and Heat

The data you obtained from Activity 4-10 support the idea or hypothesis that heat is transferred from a hotter substance to a colder one. A **hypothesis** is a set of ideas or models that a scientist puts together to try to explain how or why an event occurs in the natural world. The scientist tests the hypothesis by carrying out experiments. Other scientists repeat the experiments. If all the experiments have supported the hypothesis, and none have shown it to be false, the hypothesis is accepted as a **theory**. A theory continues to be accepted until or unless new evidence is found to disprove it.

We can use a theory about heat and temperature to improve our definition of temperature: **Temperature** is a measure of the energy level of an object or substance. It can be measured in degrees Celsius. We can define **heat** as the energy that is transferred from a hotter object or substance to a colder one.

People have not always thought of heat as energy that is transferred, as the cartoon shows.

Designing an Experiment to Test Lavoisier's Hypothesis

Problem

What happens to the mass of a sample of water when it is heated?

Define the problem: If water is heated,
(a) will its mass increase?
(b) will its mass decrease?
(c) will its mass stay the same?

Procedure

1. Predict whether the mass of the water will increase as it is heated. Write your prediction in your notebook.
2. Using the materials listed, design an experiment to test Lavoisier's hypothesis.
3. Decide how to record and organize your data.

Finding Out

1. Analyse the data you obtained. Do the data provide evidence to support your prediction?
2. Did your data support Lavoisier's hypothesis?

Materials

50 mL water   balance scale
100-mL beaker   hot plate

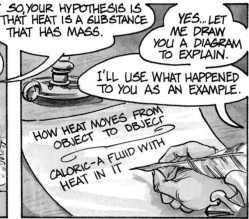

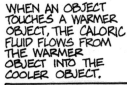

Water to Ice

Soon after Lavoisier developed his hypothesis, another scientist also investigated heat. Benjamin Thompson, also known as Count Rumford of Bavaria, measured the mass of a sample of water as it turned to ice. If Lavoisier's theory were true, the water would lose mass as it changed to ice. Thompson used very careful measurements to show that the mass did not change. He concluded that whatever "flows" from a hot object to a cold one has no mass. In other words, Lavoisier's hypothesis that a "caloric" fluid had mass could not be right. Scientists now call whatever "flows" or is transferred in this way "heat."

Benjamin Thompson, Count Rumford of Bavaria

100 g of water

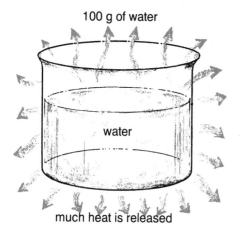

water

much heat is released

100 g of ice

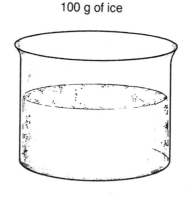

When water changes from a liquid to ice, it releases a large amount of heat. If the heat had any mass, the ice sample would have less mass than the water. However, Count Rumford proved through very careful measurements that the mass remained constant.

Heat and Changes of State

Do you remember the three states of matter—solid, liquid, and gas? If you leave an ice cube in a beaker at room temperature, heat from the surrounding air will transfer to the ice, and the ice will become water. When ice melts, it changes from a solid to a liquid. **Melting** is a change from one state to another. If you leave the water standing in the beaker for several days, it will disappear, or evaporate. When water evaporates or boils, it goes from a liquid to a gas. Thus, **evaporation** is another example of changing from one state to another.

When a substance changes state, what happens to its temperature? Do the next two Activities to answer this question.

Ice to Water

Problem

What happens to the temperature of ice as heat is transferred to it from the surrounding air?

Materials

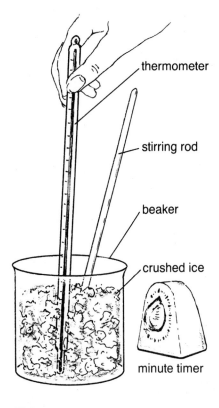

thermometer

stirring rod

beaker

crushed ice

minute timer

Procedure

1. In your notebook draw a table similar to Table 5. Continue the time column to 12 min.
2. Stir the crushed ice with a stirring rod. Place the thermometer in the beaker and record the temperature at 0 min. Take out the thermometer and stir the ice some more. Again insert the thermometer in the beaker and measure and record the temperature. Record the temperature every minute. In the right-hand column, describe any changes you observe.

> **CAUTION:** Do not stir the ice while the thermometer is in the beaker.

3. Continue for 12 min or until all the ice has melted, whichever comes first.

Finding Out

1. What change of state occurs in this Activity?
2. This change of state needs energy. Where did this energy come from? What do we call this energy?
3. What happens to the temperature during this change of state?
4. On the basis of your observations from this Activity, do you agree with the following statement: "When heat is added to a solid, the heat can cause either a change of state or an increase in temperature"? Explain your answer.

Finding Out More

5. Draw a line graph of your data. Describe the shape of the line during the change of state called melting.
6. Predict what would happen to the line on the graph if you heated the water after all the ice had melted.

Table 5

TIME (min)	TEMPERATURE (°C)	CHANGES OBSERVED
0		
1		
2		

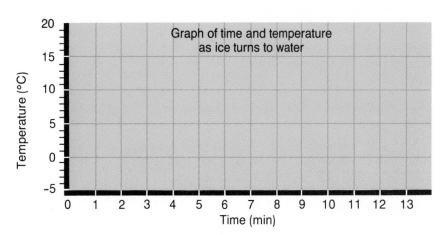

Graph of time and temperature as ice turns to water

Water to Steam: Demonstration

Problem

What happens to the temperature of water as it boils?

Materials

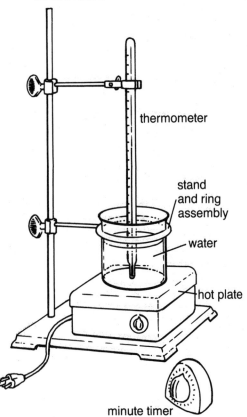

thermometer

stand and ring assembly

water

hot plate

minute timer

Note: Use a thermometer that is accurate at temperatures above 90°C.

Procedure

1. In your notebook, draw a table to record your data similar to the one used in the previous Activity. The time column in this table should go from 0 min to at least 15 min.

CAUTION: Be careful when working with boiling water. Do not put your hand in the water or in the steam.

2. Make a prediction about what happens to the temperature of water as it boils.
3. Add about 100 mL of hot water from the tap to the beaker. Begin heating the water.
4. When the temperature reaches about 60°C, record your first temperature reading at 0 min.
5. Measure and record the temperature every minute.
6. In the right-hand column of your table describe any changes you observe.
7. Continue recording the temperature for at least 5 min after the water has been boiling rapidly.

Finding Out

1. What change of state occurs in this Activity?

2. What happens to the temperature during this change of state?
3. Was your prediction correct? Explain.
4. On the basis of your observations from this Activity, do you agree with the following statement: "When heat is added to a liquid, the heat can cause either a change of state or an increase in temperature"? Explain your answer.

Finding Out More

5. Draw a line graph of your data. Describe the shape of the line
   (a) before the water boiled;
   (b) while the water was boiling.

Extension

6. Do you think the boiling temperature of water changes for different amounts of water? Design an experiment to find out. Check your design with your teacher before conducting your experiment.

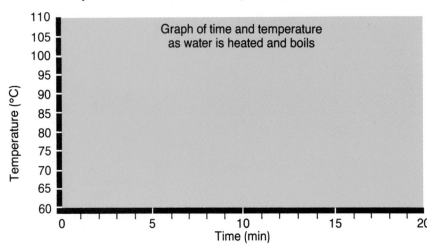

Graph of time and temperature as water is heated and boils

Temperature (°C): 110, 105, 100, 95, 90, 85, 80, 75, 70, 65, 60

Time (min): 0, 5, 10, 15, 20

Heat and the Particle Theory

Did You Know?

Some people say that if you need ice cubes in a hurry, you should fill your ice-cube tray with hot water, not cold. Sounds like crazy advice? Well, it is and it isn't. Suppose you put hot water in one ice-cube tray and an equal mass of cold water in another ice-cube tray, then the hot water *will* freeze faster. Why? Because some of the hot water will change into gas in the freezer. Then the mass of the remaining hot water will be less than the mass of the cold water, and the hot water will freeze faster. Try it for yourself, but be sure to use equal masses of hot and cold water.

And if you really need ice cubes in a hurry? Try only *half-filling* the ice-cube tray with cold water.

What exactly is the difference between water at 10°C and water at 20°C? How are ice and water different? Are you still wondering exactly what heat and temperature are?

Scientists over the years have thought of many ideas, or hypotheses, to try to explain heat and temperature. One such hypothesis was thought of by Lavoisier. As you have read, he suggested that heat might be a substance with mass, which he called caloric. But Lavoisier's idea was not supported by the observations you made in Activity 4-11, and those of other scientists who were looking for ways to explain heat.

Scientists now use the **particle theory** to explain heat and temperature. This theory is based on a model that suggests that all matter is made up of tiny particles too small to be seen. According to the model, these particles are always moving—that is, they have energy. The more energy they have, the faster they move. For example, both hot and cold water are made up of moving particles. But the particles move faster in hot water than in cold water.

So far, all the evidence that scientists have about matter supports the idea that all substances are made up of moving particles. That is why we call the particle model for matter a theory.

Did You Know?

If we could somehow line up 10 million hydrogen particles in a straight line, the line would be only 1 mm long.

A Model of Moving Particles

Problem

What happens to the volume that particles occupy when they change speed?

Materials

flat open box with sides about 10 cm to 20 cm high

about 30 to 60 sheets of used writing paper (the number will depend on the size of the box)

Procedure

1. Crumple the sheets of paper into tight balls. Each paper ball will represent a particle.
2. Empty the box. Choose two students to be the "shakers." Other students holding the "particles" will be gathered near the shakers.
3. The two shakers grasp the ends of the box, and push it back and forth between them (not up and down) very quickly. Add one particle at a time to the box. Find the greatest number of particles the box will hold at high speed.
4. Repeat Step 3 at medium speed, and find the number of particles the box will hold.
5. Repeat Step 3 at slow speed, and record the number of particles.

Finding Out

1. Which require more space, slow-moving or fast-moving particles?
2. Does a particle collide with other particles more often at low speed or at high speed?
3. Imagine that the "particles" used in this Activity represent the particles of a gas, such as air. Which step represents gas at a low temperature? at a high temperature?
4. Use the particle theory to explain why a gas expands when it is heated.
5. The particle theory applies to all states of matter. Do you think the paper ball "particles" could represent particles of a liquid? Explain.
6. Which diagram best represents the motion of one particle of a gas at a high speed?

Finding Out More

7. Use the particle theory to explain how a gas thermometer works. What happens in the thermometer as the temperature goes up?
8. Use the particle theory to explain the behaviour of a liquid such as alcohol in a liquid thermometer.

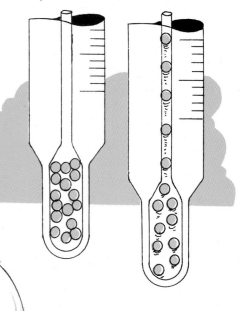

(a)

(b)

(c)

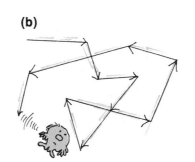

Probing

1. What design features make kettles more efficient for heating water than saucepans?
2. If you didn't have a kettle, what would be the most efficient type of pan to use to heat water quickly?

Temperature and the Particle Theory

Some individual particles of a substance may have a lot of energy and move very quickly. Other particles of the substance may have less energy. But in any substance, the particles have an **average energy**. When we talk of temperature as a measure of the energy level of an object or substance, we mean that it is a measure of this average energy of the particles.

Expansion, Contraction, and the Particle Theory

The particle theory is a useful model to explain why substances expand when they are heated and contract when they are cooled. At high temperatures, particles have more energy, move more quickly, and have more collisions. As a result, they take up more space: their volume increases—they expand. At lower temperatures, particles have less energy, move more slowly, and have fewer collisions. Thus, they take up less space: their volume decreases—they contract.

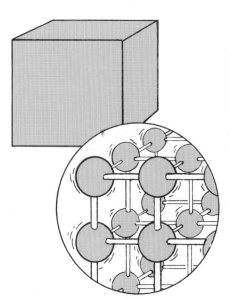

Particles in a solid move, but only by vibrating in the same spot.

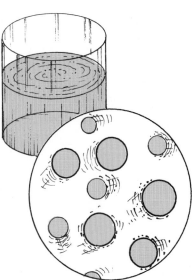

Particles in a liquid are free to move around.

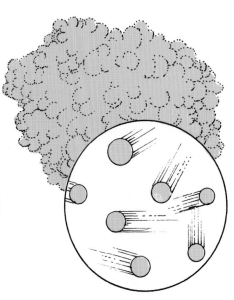

Particles in a gas move independently and are separated by large spaces.

Changes of State and the Particle Theory

The particle theory can also explain what happens in a change of state. You found out that adding heat to ice at 0°C did not cause much change in the temperature—it simply changed the ice to water. In other words, when a solid changes to a liquid, the heat being transferred to the solid is being used to bring about the change of state. Similarly, when liquid changes to a gas—for example, when water boils—the temperature remains about the same, and the heat is being used to bring about the change of state.

In the next Activity you will look more closely at what happens when liquid changes to a gas—the change of state called evaporation.

After you get out of a lake or swimming pool, water evaporates from your skin. The evaporation makes your skin feel cooler. Does that mean the evaporation is taking heat away from your body? Test this by evaporating water from a thermometer and observing what happens to the temperature.

Heat and Evaporation

Problem

When water evaporates from the surface of an object, what is the effect on the temperature of the object?

Materials

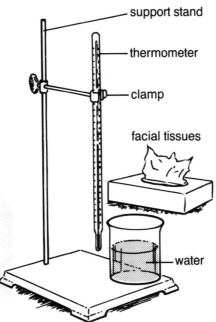

support stand

thermometer

clamp

facial tissues

water

Procedure

1. Predict whether the temperature of an object goes up, goes down, or stays the same when water in contact with the object evaporates. Give a reason for your prediction.
2. Take a small piece of facial tissue. Dampen it, and wrap it around the bulb of the thermometer.
3. Observe and record the temperature for about 8 to 10 min.

Finding Out

1. What was the final temperature you recorded?
2. Where did the energy for evaporation come from? How do you know?

Finding Out More

3. If a fan is available, use it to blow air on the bulb of the thermometer. Dampen the tissue again before you do so. Observe any changes in temperature, and explain them.
4. Use the particle theory to explain why the temperature of the thermometer changed.

Energy is transferred from the thermometer to the water particles.

A particle of evaporated water has more energy than a particle of liquid water.

Extension

5. Put a small amount of rubbing alcohol on the back of your hand. What happens as the alcohol evaporates? Explain your observation.

Making a Hypothesis and Testing It

In the next Activity you will be asked to make a hypothesis, and you will design an experiment to test your hypothesis. Recall that scientists make hypotheses, which are ideas about relationships and patterns. From these hypotheses, they predict the outcome or behaviour of something. Then they carry out experiments to find out whether their predictions were correct and there is evidence to support their hypotheses.

Activity 4-16

How Fast Can Water Evaporate?

Problem

How can you evaporate one litre of water at room temperature as quickly as possible? (You may use three different one-litre samples of water in various containers or in various situations. However, you may *not* use any type of heat source in this experiment.)

Materials

You will be able to list the materials you need only after you have formed your hypothesis and designed your experiment.

Procedure

1. First form your hypothesis: What is the factor or factors that you think will influence evaporation of water?

2. In your notebook, design an experiment to test your hypothesis. Predict the fastest way to evaporate the water, based on your hypothesis.
3. Decide how you can use other samples of water to check your hypothesis.
4. Before you begin, decide
   (a) how you will obtain the data from your experiment;
   (b) how you will record and organize your data;
   (c) how you will analyse and interpret your data.

Finding Out

1. Did your results support your predictions based on your hypothesis?
2. Have a class discussion about the various methods. Which was the fastest method of evaporation? Why?

Finding Out More

3. Which is more likely to dry up in hot summer weather, a body of water that is 30 m long, 30 m wide, and 40 m deep, or one that is 60 m long, 60 m wide, and 10 m deep?
4. Why should you not wear damp clothes on a cold day?

Extension

5. If one is available, examine a drum-type humidifier. Explain:
   (a) why a foam rubber pad is wrapped around the drum;
   (b) why a fan is used in the humidifier.

Temperature, Total Energy, and Heat: a Summary

Study the three situations shown to be sure you understand how temperature, total energy, and heat compare to each other.

Two samples of water at 50°C.
The temperatures are the same because the average energy of the particles is the same. The total energy is greater in the larger sample because the larger sample contains more particles with the same amount of energy. If the two samples are mixed, no heat is transferred.

Two equal amounts of water at different temperatures.
The temperature is higher in the warmer water because the average energy of the particles is greater. The total energy is greater in the warmer water. If the two samples are mixed, heat transfers from the warmer to the cooler water.

Two unequal amounts of water at different temperatures.
The temperature in the smaller sample is higher because the average energy is greater. The total energy cannot be compared. If the two samples are mixed, heat transfers from the smaller sample to the larger one.

Science and Technology in Society
A Cold and Courageous Story

Have you ever been really cold, maybe by getting soaked right through during an icy blizzard? Sometimes, people may become so cold that they die or are near death. When their body temperature drops to dangerously low levels, they are in a condition known as *hypothermia*.

When you are cold and uncomfortable, you try to get warm by putting on more clothes or adjusting the thermostat. But what if you are a marathon swimmer, who spends many hours, even days, in cold water?

Marathon swimmers need to know about heat and temperature. Their swimming ability, even their lives may depend on an understanding of heat, and the effects cold water can have on body temperature.

In the summer of 1988, Vicki Keith swam across all five Great Lakes. She was in Lake Michigan for 53 hours, and in Lake Huron for 47 hours. When she swam across Lake Superior, the water temperature was as low as 11°C.

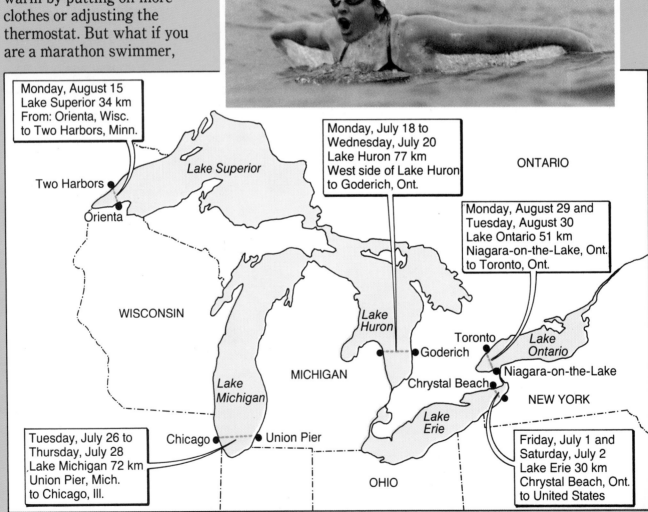

Monday, August 15
Lake Superior 34 km
From: Orienta, Wisc.
to Two Harbors, Minn.

Monday, July 18 to
Wednesday, July 20
Lake Huron 77 km
West side of Lake Huron
to Goderich, Ont.

Monday, August 29 and
Tuesday, August 30
Lake Ontario 51 km
Niagara-on-the-Lake, Ont.
to Toronto, Ont.

Tuesday, July 26 to
Thursday, July 28
Lake Michigan 72 km
Union Pier, Mich.
to Chicago, Ill.

Friday, July 1 and
Saturday, July 2
Lake Erie 30 km
Chrystal Beach, Ont.
to United States

Two Harbors
Orienta
Lake Superior
ONTARIO
WISCONSIN
MICHIGAN
Lake Michigan
Lake Huron
Chicago
Union Pier
Goderich
Toronto
Lake Ontario
Niagara-on-the-Lake
Chrystal Beach
NEW YORK
Lake Erie
OHIO

When Vicki planned her long-distance swims, she knew that the cold would be as much of a challenge as the distance she had to swim, so she took the following precautions.

1. Before each swim, she deliberately gained 10 kg by eating huge amounts of foods that have a high carbohydrate content. This extra covering of fat helped protect her body from losing heat to the cooler water. It also served as stored energy for her gruelling marathons. During each swim Vicki lost 10 to 14 kg.

2. Every two hours during her swims, Vicki ate high-energy foods—hot soup, hot chocolate, chocolate bars, and fruit cocktails.

Vicki says, "It was noticeable that I wasn't swimming as well at the end of two hours. I'd be cold and tired because I'd run out of fuel."

3. Vicki applied a lanolin ointment all over her body, paying special attention to the areas that would lose the most heat. This ointment was water-repellant, and prevented the cold water from touching her skin. She also wore two bathing caps to slow down the heat loss from her head. The pocket of warm air that formed between the caps helped to insulate her head further.

4. By wearing a black bathing suit, Vicki was able to absorb a little heat from the sun during the daytime. The nights seemed particularly cold, and her body shivered continuously, generating more heat.

Think About It
Sometimes Vicki would hit a warm patch of water in the lake. What a relief! Her crew travelling in the boat beside her, however, thought the water was still very cold. Why did the water seem warm to Vicki?

Checkpoint

1. In your notebook, copy the words in the left-hand column below. Beside each word write the best matching description from the right-hand column.

(a) thermostat	A temperature-measuring device that contains two different metals touching each other at one end.
(b) thermocouple	This device is likely to contain a bimetallic strip.
(c) temperature	A measure of the average energy of the particles of a substance.
(d) heat	
(e) thermograph	This device can be used to detect weak points in your home insulation.
	The energy transferred from one substance to a colder one.

2. Two cups contain the same amount of water.
   (a) What device would be best for comparing the average energy of the particles of each sample of water?
   (b) Explain how this device would help you compare the energies.
3. Describe the ways in which a clinical thermometer and a laboratory thermometer are different.
4. Use the particle theory to explain why solids contract when cooled.
5. Jenny was drying her hair when suddenly the dryer stopped working. Jenny wasn't worried; she just turned the switch to OFF and waited until the dryer cooled. Then she turned it on and it worked again. Jenny knew that a thermostat controlled overheating in her hairdryer. Describe how it might work. Use a diagram in your answer.
6. What happens to the average energy of the particles in milk when the milk is taken from the refrigerator and begins to warm up?

7. Think of a situation where a solid thermometer would be more useful than a liquid one.
8. Anders is trying to make some popsicles from fruit juice in the freezer. He wants to eat them as soon as possible, so he puts a thermometer in one. Every ten minutes he checks the temperature of his popsicles. At first, everything is going well, but suddenly his popsicles stop getting colder! After a long wait, though, they start getting colder again, and he sees they are hard enough to eat.
   (a) Look at the temperature readings Anders obtained. Make a line graph of Anders' data.
   (b) At what temperature does a flat region appear on the graph? What do you think was happening at this time?
   (c) At what temperature do you infer that the fruit juice froze? Explain your answer.

Temperature Readings

TIME (min)	TEMPERATURE (°C)
0	20
10	14
20	8
30	2
40	-2
50	-2
60	-2
70	-4
80	-10
90	-12
100	-12

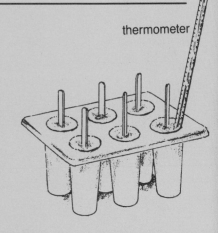

thermometer

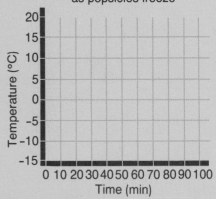

Graph of time and temperature as popsicles freeze

Measuring Heat

Heat is a form of energy. Energy is measured using a unit called the **joule** (named after James Joule, an English scientist who studied heat and other forms of energy). The symbol for joule is J. One joule is a small amount of energy. Often it is more useful to measure energy in kilojoules (1 kJ = 1000 J) or megajoules (1 MJ = 1 000 000 J).

To get an idea of the size of these units of energy consider these facts:

- It takes about 1 J of energy to lift an average-sized hamburger from the table to your mouth.
- It takes about 1 kJ of energy for you to climb halfway up a flight of stairs.
- It takes about 1 MJ of energy from gasoline to drive a compact car about 300 m.

You found in Unit One that all living things need food to survive. The food provides energy for them to move, grow, and reproduce. Your body uses the food you eat to produce the energy you need to swim, hike, run, lift your arm, wriggle your toe—every single action you take. The energy stored in food also produces heat to keep your body temperature at about 37°C. You have read that food contains stored energy, but how can you be sure? What evidence can you obtain that shows food can release energy? Heat is a form of energy. Does food release heat?

Energy from Food

PART A

Problem

What evidence shows that energy is stored in a food such as a peanut?

Materials

peanut
pin embedded in cork
Pyrex test tube
prepared large juice can
10-mL graduated cylinder
10 mL of water
thermometer
matches
safety glasses

Procedure

1. Think about these questions:

HOW IS FOOD SIMILAR TO FUEL SUCH AS GASOLINE?

CAN THE ENERGY OF A BURNING FUEL BE USED TO HEAT WATER?

WHAT HAPPENS TO THE TEMPERATURE WHEN YOU APPLY MORE HEAT TO A SAMPLE OF WATER?

HOW CAN YOU MEASURE HEAT?

2. Predict what you think will happen if you try to heat water by burning a food such as a peanut. Give reasons for your prediction.

3. Very carefully place a peanut on the sharp end of a pin that has been embedded in a cork as shown in the diagram.

4. Place a Pyrex test tube in a hole in the bottom of an upside-down can. (Make sure the test tube fits properly so that it remains suspended as shown.) Now slide the cork and peanut under the test tube. There should be a space of about 1 cm between the test tube and the peanut. If necessary adjust the height of the test tube in the can.

5. Measure 10 mL of water, pour it into the test tube, and place a thermometer in the test tube. Measure the temperature of the water.

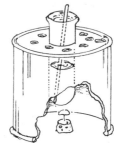

CAUTION: Be careful when using a flame.

6. Now you're ready to begin your experiment. Light the peanut with a match. As soon as you are sure it is burning, place it directly under the test tube within the can. Let the peanut burn completely, then immediately record the final temperature of the water.

PART B

Now design an experiment to compare the energy produced by a walnut and a cereal (such as a Shreddie) with the peanut. Take care to make your comparison a fair test.

Finding Out

1. Do your results support your prediction?
2. Do you think any energy from the food was not measured in the Activity? If so, what happened to it?
3. How could you improve the experimental design to measure the energy produced by the burning food?

Finding Out More

4. Describe what the peanut looked like after you finished the investigation.

Energy Content of Foods

How is knowing about the energy content of food important to you? Think about this fact. A chocolate milkshake has 4 950 J in every gram. A typical milkshake has a mass of about 400 g. It will release a total of 1 980 000 J (1 980 kJ) of energy for use by your body. For someone your age, it would take almost two hours of bicycling to use up this energy. Table 6 compares chocolate milkshakes to several other common foods.

Probing

Use Table 6 to answer these questions:
1. Which food has the highest energy content per serving?
2. Which food has the highest energy content per gram?
3. Which has the higher energy content per gram—chicken or beef?
4. How many hours and minutes would you have to bicycle to use up the energy of a chocolate bar and a slice of apple pie?

Table 6

FOOD	MASS OF TYPICAL SERVING IN GRAMS (g)	ENERGY CONTENT IN JOULES PER GRAM (J/g)	ENERGY/ SERVING IN KILOJOULES (kJ)	TIME OF BICYCLING TO USE UP THIS ENERGY IN MINUTES (min)
Chocolate milkshake	400	4 950	1 980	110
Fried egg	46	7 350	338	19
Wiener	37	3 200	118	6.5
Ground beef	88	12 000	1 074	60
Fried chicken	140	2 600	364	20
Peanut butter	16	24 750	396	22
Banana	114	3 850	439	24
Apple pie (slice)	158	10 700	1 692	94
Chocolate bar	30	19 600	589	33

Energy Stored in Fuel

Fuels are used to provide heat for cooking, for heating buildings, and for making products in many industries. In the past wood and coal were the most common fuels. Today fuels made from crude oil and natural gas are more widely used than wood and coal.

Comparing Various Fuels

The energy stored in fuels can be used to heat homes directly, or to operate cars, trucks, and planes. It can also be used to produce electricity. Can you think of other uses? The table shows how much energy can be obtained from some of our important fuels.

The Energy Content of Burner Fuel: Demonstration

Problem

What is the energy content of burner fuel?

Probing

Compare these values with the ones for food on page 237. Make a bar graph to illustrate the differences. On your graph, place 4 fuels and 4 foods on the x-axis.

1. Which food contains almost as much energy per gram as a fuel that we burn?
2. Why don't we use this food as a fuel?

Table 7 *Energy Content of Fuels*

FUEL	ENERGY CONTENT IN JOULES PER GRAM (J/g)
coal (solid)	28 000
gasoline (liquid)	44 000
kerosene (liquid)	43 000
methane (gas)	49 000
ethane (gas)	44 000
propane (gas)	43 000

Materials

burner
burner fuel
400-mL beaker
thermometer
stirring rod
balance scale

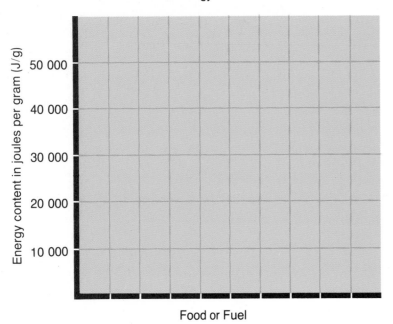

Comparison of energy from food and energy from fuels

Procedure

CAUTION: Before you start, make sure you know how to turn the burner off.

1. Add a safe amount of fuel to the burner and determine the mass of the burner and the fuel.
2. Add 200 mL (200 g) water to the beaker and place it above the burner.
3. Make a table in your notebook similar to Table 8. Record the mass of the burner and fuel, and the temperature of the water at the beginning of the demonstration.
4. Light the burner and stir the water while it is heating. When the temperature reaches between 80°C and 90°C, turn the burner off.
5. Record the highest temperature reached.
6. Record the final mass of the burner and fuel.

Table 8

	TEMPERATURE OF WATER (°C)	MASS OF BURNER AND FUEL (g)
Start		
Finish		
Difference		

Finding Out

1. Follow these instructions to calculate the energy that the fuel transferred to the water.
   (a) Subtract the starting temperature from the final temperature to find the change in temperature.
   (b) Subtract the final mass of the burner and fuel from the starting mass of burner and fuel to find the change in mass.
   (c) Do this calculation:

$$\frac{\text{temperature change of water} \times 840}{\text{change in mass of burner and fuel}}$$

   State your answer in joules per gram (J/g)

2. The energy content of most liquid fuels is about 40 000 J for each gram of fuel. Compare this value with your answer in 1(c). Explain why your result might be different.

A LET'S SEE, WE NEED TO SUBTRACT THE STARTING TEMPERATURE OF 15°C FROM THE FINAL TEMPERATURE OF 85°C. THAT GIVES US 70°C, SO THE FUEL HEATED THE WATER BY 70°C. [85°C − 15°C = 70°C]

B THE STARTING MASS OF FUEL WAS 54.5 g. WE ENDED UP WITH A FUEL MASS OF 51.5 g. WE USED 3 g. OF FUEL TO HEAT THE WATER. [54.5 g − 51.5 g. = 3.0 g]

C TO DO THE CALCULATION, FIRST WE MULTIPLY THE TEMPERATURE INCREASE, 70°C, BY 840 — — THAT GIVES 58 800 [70 × 840 = 58 800] NOW WE DIVIDE 58 800 BY THE DIFFERENCE IN FUEL MASS, 3 g — — THAT IS, 19 600. $\frac{58\,800}{3} = 19\,600$ THERE IS 19 600 UNITS OF ENERGY IN EVERY GRAM OF FUEL, OR 19 600 J/g.

Sources of Heat

There is a lot of heat available in fuels—solid fuels, such as coal; liquid fuels, such as oil; and gaseous fuels, such as natural gas. Western Canada is especially well supplied with fuels that can be used to produce heat: oil, natural gas, and the mixture that forms the tar sands. How do these substances produce heat?

Chemical Energy as a Source of Heat

Food and fuels have energy stored in them. This kind of energy is called **chemical energy**. Heat is released from food and fuels through the chemical process of burning. To release the heat in your body, food is digested instead of being burned. The digestive process releases the heat gradually so that there is no damage to your body.

As the wood burned, energy stored in the wood provided the heat needed to roast these hot dogs.

Many electrical devices produce heat even though their main purpose is something else. An ordinary light bulb uses about 5% of its electrical energy to produce light. It gives off about 95% of its energy as heat.

Electrical Energy as a Source of Heat

Electricity can be used as a source of heat, but we must produce the electricity first. There is more than one way to produce electricity. In Canada we are lucky to have many large river systems. The energy in their moving water can be used to generate electricity. Electricity produced by this process is called **hydroelectricity** ("hydro" means water). Much of our electricity is also produced at generating stations that burn fuels such as coal to produce electricity.

Once the electricity has been produced, it can be sent through wires to homes and industries. In our homes we use it as a source of heat in stoves, toasters, dryers, heaters, and many other appliances.

One disadvantage of generating stations that burn fuels such as coal is that a great deal of heated water is released into nearby waterways, threatening living things in those environments. Another disadvantage is that certain chemicals are released from smokestacks when the fuels are burned. These chemicals can harm the environment.

Just by examining these two photos, can you suggest two advantages of hydro-electric over coal-fired generating stations?

Mechanical Forces as a Source of Heat

Do you remember the definition of temperature? It is a measure of the average energy of the particles of a substance. If the particles in a substance move more quickly, the average energy increases, and the substance gets hotter.

Unit Three described how a force is a push or a pull made directly on an object. You can use the push or pull of a mechanical force to make the particles of a substance move more quickly. You used the mechanical force called friction in Activity 3-11 when you rubbed a nail against a brick and noticed that heat was produced. In the next Activity you will investigate this and other ways of using mechanical forces to produce heat.

Activity 4-19

Heat from Mechanical Sources

Problem

What are some ways of producing heat using mechanical means?

Materials and Procedure

1. Repeat the friction test you did in Activity 3-11 using a paperclip and taking these steps.

   (a) Straighten a paperclip and test its temperature by touching it to the palm of your hand.

   (b) Rub the paperclip vigorously with your fingers for at least 10 s.

   (c) Very quickly test the temperature with the palm of your hand again. Write your observations in your notebook.

2. (a) Straighten a paperclip and bend it back and forth quickly about 10 times.

   (b) Test the temperature of the region where you bent it, again by touching it to the palm of your hand.

   (c) Describe what you feel in your notebook.

3. (a) Test the temperature of a wooden block by touching it with the palm of your hand.

   (b) Pound one place on the block about 10 times with a hammer.

   (c) Test the temperature of the wood where you pounded it.

   (d) Describe what you feel. For this test, you could use a piece of metal instead of a wooden block.

CAUTION: Be careful when working with sharp objects.

Geothermal Energy as a Source of Heat

The inside of the Earth is very hot. Sometimes the hot materials from inside the Earth find a way through to the surface. Then we see a dramatic result such as an erupting volcano or an active geyser. The energy inside the Earth is called **geothermal energy**.

In some parts of the world, people use geothermal energy to heat their homes or to produce electricity. This is done in Iceland, New Zealand, Italy, and the northwestern United States. In Canada, there are hot springs in British Columbia and Alberta that we could use as sources of heat. However, because we have so many other sources of heat, we are not yet using these sources of geothermal energy.

4. (a) Feel the valve of a bicycle pump.
(b) Compress the air in the pump, then feel the valve again.
(c) Describe what happened.

Finding Out

1. Based on the steps in this Activity, what are four ways of producing heat by mechanical means?
2. Use the particle theory to explain why the temperature of an object can increase because of friction.
3. Explain how heat is produced in each of the following cases:
   (a) a basketball player trips and skids along the gymnasium floor;
   (b) a carpenter uses a hammer to force a nail into a wooden board;
   (c) during baseball practice, a catcher's hand feels warm after several fast pitches.

Solar Energy as a Source of Heat

Build a model solar hot-water heater. How can you use the sun's rays to heat water? Think about it, then design and build a solar heater of your choice!

Anyone who has been outside on a summer day knows that the Sun produces heat! The Sun is the source of **solar energy**. Is there any way of capturing this energy and using it at night or on a cloudy day?

In some areas, homes can be specially designed to use solar energy for heating living areas. These homes need large, south-facing windows, and thick walls and floors made of concrete. As sunlight enters the windows, it heats both the air inside the house and the concrete walls and floors. During the night, the warm concrete releases its heat to the inside of the house. The windows are covered so that heat does not escape.

Solar energy can also be used to heat water in devices called solar collectors. Solar collectors are usually placed at a carefully calculated angle on the south-facing roof of a building. Some collectors are designed to use the hot water to heat the inside of the building; others simply heat the water for use as hot water.

Diagram of a solar heated house, showing thick walls and large, south-facing windows.

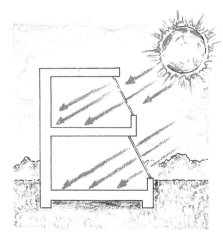

Solar heated house during the day.

shutters closed

Solar heated house at night.

Solar collectors heat about one-third of the hot water for this retirement home in Port Alberni, British Columbia. This project has shown that solar radiation is a valuable energy resource even in cloudy areas.

This solar furnace uses a lens to focus light on a fireproof brick.

The plant and animal material in garbage can be burned to produce heat. Garbage can also be used to make fuels called methane and methanol. As our supplies of oil and natural gas become more scarce and therefore more expensive, garbage may become a more common source of heat.

Thinking about Sources of Heat

Renewable energy sources are those that are not destroyed or used up in the process of being used—electricity produced from the moving waters of rivers and energy from the Sun are two examples. **Non-renewable energy** sources are gradually being used up, and one day these sources may run out entirely —examples are coal and oil. In the three million or so years that humans have been on Earth, we have produced heat mainly from renewable fuels such as wood. In the last few hundred years, people have used non-renewable energy resources at an increasing rate.

There is a danger that in the future we could use up all the sources of non-renewable energy. We may need to switch to using renewable sources. Every member of our society may have to make difficult decisions about which sources of energy to use. We certainly can't take energy sources for granted.

This experimental car uses solar energy. The car cannot be driven at night or on cloudy days. It is too light and too fragile to carry more than one person or to be driven in traffic.

Probing

1. As you have read, solar energy can be used to produce heat. Solar energy is an example of radiant energy. Visible light, microwaves, infrared light, ultraviolet light, and X-rays are all examples of radiant energy. Select one of these other forms of radiant energy, and investigate its main uses.

2. Nuclear energy can also be used as a source of heat. Find out how heat is produced using nuclear energy. Where is the heat produced from nuclear energy used? What are some of the advantages and disadvantages of producing nuclear energy?

Working by Helping People

Premature babies must often be rushed by air ambulance to a distant hospital if they are to survive. Thanks to Dr. John Smith, a biomedical engineer, these tiny babies now have a better chance of living through the journey.

The portable incubators used in the air ambulances are heated by forced air flowing through the incubator. The forced air in the incubators used to flow so quickly that it caused the baby to lose body heat. A baby's skin is very thin. Consequently heat was easily transferred from its body to the air. Dr. Smith has redesigned the incubator so that special channels circulate the air at slow speed, reducing the dangerous loss of body heat.

As a child, Dr. Smith liked to spend time making things, fixing car engines, and "pulling things to bits." When he went to university, he decided to study engineering. "Engineering is a broad-based education that gives you an opportunity to follow many careers. I think it's important that students get a solid background in sciences, and then they can jump off in any direction."

Dr. Smith first took a degree in electrical engineering and then specialized in biomedical engineering.

Some biomedical engineers work in research laboratories; Dr. Smith works in a hospital. "At any moment a small child may wander into my office or a doctor may stop me in the hall to talk about improving a medical device. Then I realize how biomedical engineers have a direct role in improving life for the patients." Dr. Smith has developed more than ten medical engineering devices that are helping to save the lives of babies and children.

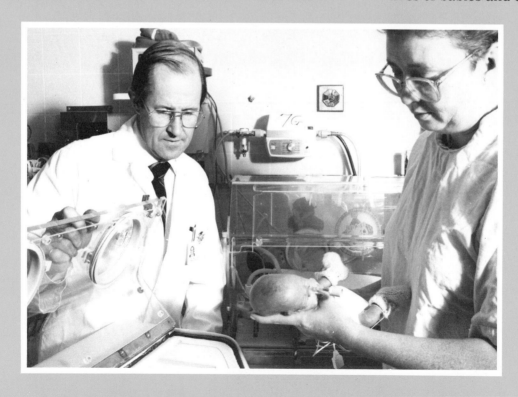

Checkpoint

1. (a) Unscramble each of the following terms.
   LUJOE
   HINACLEAMC REFCO
   CILMEACH REGENCY
   DROLYHETRECCITIY
   ROLAS GREENY
   MOTHERGEAL YERNEG
   (b) Use two of the terms in sentences that illustrate their meaning.

2. In each situation, state the type of energy that is the source of heat.
   (a) The blade of an electric saw becomes hot as it cuts through a log.
   (b) An iron is used to press clothes.
   (c) A match bursts into flame.
   (d) A water heater in the home provides hot water for washing.
   (e) A greenhouse produces tropical plants.
   (f) A toaster is used in preparing breakfast.
   (g) A nail becomes warm as it is pounded into a board.
   (h) The engine of a gasoline mower is operated without oil, so it overheats.

3. (a) What unit do we use to measure energy?
   (b) How do we state the energy content of foods and fuels?

4. Read each of the following statements and explain why each is a misunderstanding of what you studied in Topic Five.
   (a) Food is energy.
   (b) The energy in a serving of food is measured in grams per kilojoule.
   (c) It takes about 10 kJ to lift a glass of milk from the table to your mouth.
   (d) Wood, oil, and natural gas are common forms of energy used to heat homes.

5. In a science lab, Patrick is trying to compare the amounts of heat stored in a walnut and a pecan. Look carefully at the illustration. What is Patrick doing that is unsafe? (Refer to the safety rules at the front of the book.)

6. (a) Make a list of appliances in your home that use electrical energy. Make a second list of devices that use energy from fuels.

   (b) Pick one device from each list. For each, design a device with a similar function that uses the other source of energy. (For example, design an oil-powered hairdryer or an electrical lawn mower.) Describe any major problems you might encounter in using that particular source of energy for your device.

7. Write a science fiction story about a technological breakthrough in which a totally new and cheap source of heat is discovered. How would it affect business and industry?

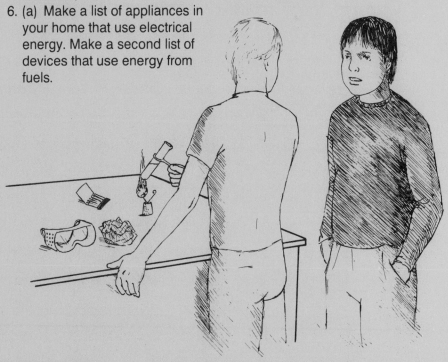

Focus

- The three states of matter are solid, liquid, and gas.
- Solids, liquids, and gases expand when heated and contract when cooled.
- A thermometer is a device used to measure temperature.
- Expansion and contraction are involved in measuring temperature.
- Heat and temperature are different: heat is the amount of energy transferred; temperature is a measure of the average energy of the particles of a substance.
- Heat can be transferred from a warmer substance to a cooler one.
- The particle theory is useful in explaining the effect of heat on solids, liquids, and gases.
- During the period when a substance undergoes a change of state, its temperature does not change.
- The energy content of foods and fuels can be measured and stated in joules per gram.
- Sources of heat include chemical, mechanical, electrical, solar, and geothermal.

Backtrack

1. Describe the difference between expansion and contraction. When does each occur?
2. State one useful application for each of the observations listed below.
   (a) Liquid mercury expands when heated and contracts when cooled.
   (b) Metal expands more than glass does when heated.
   (c) Some metals expand more than others when heated.
3. A boy who feels ill touches his hand to his forehead to see if he has a fever. Will he be able to tell? Explain.
4. Explain the significance of these temperatures:
   (a) 100°C   (b) 37°C
   (c) 20°C   (d) 0°C
5. Explain why a liquid clinical thermometer
   (a) has a small range of Celsius degrees;
   (b) has a narrow point in the bore just above the bulb.
6. Describe the difference between a bimetallic strip and a thermocouple.
7. State the type of energy that is the source of heat in each situation described below.
   (a) A dentist's high-speed drill becomes hot when drilling teeth.
   (b) We keep our bodies at a temperature of about 37°C.
   (c) Ancient people used flint, a very hard type of stone, to start fires.
   (d) Water at Banff Hot Springs is heated naturally.
   (e) Water from the Earth's oceans and lakes evaporates, forming clouds.
8. Explain why gaps have been left in the concrete top of the bridge in the photograph.

Synthesizer

9. Many people turn down the heat in their homes before going to bed in the winter. Then they often hear the floors start to creak. Why does this creaking occur?

10. Explain why an ordinary glass bowl taken from the refrigerator and put into a hot oven is likely to crack, but Pyrex probably won't.

11. Masahiro has an electric heater in her room at home. She wants the heater to turn on whenever the room temperature drops to 19°C. Design a thermostat using a bimetallic strip to solve this problem. Your design should include a diagram.

12. Steel rods are used in concrete buildings to help strengthen the concrete. Steel and concrete expand the same amount when heated. Describe what might happen if they didn't.

13. Some vegetables are being cooked in a pot on a stove. The lid starts to jump up and down. Use the particle theory to explain why this happens.

14. Explain, in terms of particles, what happens when a liquid changes into a gas. Use the idea of heat in your explanation.

15. Explain why a liquid clinical thermometer has a very fine bore.

16. Describe how a bimetallic strip might be used to make a thermometer. How would you make a scale for this thermometer? Would the thermometer be very accurate? Explain.

17. Look at the experimental set-up. The water in both beakers starts off at the same temperature and the hot plates produce the same amount of energy. In which beaker will the water boil first? Explain, using the words "particles," "average energy," and "heat" in your explanation.

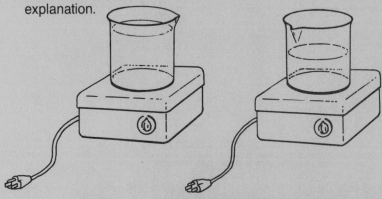

18. Newspapers often refer to solar energy as "free." Is it really "free"? Explain what expenses you would have in converting your home to solar heating.

19. Using what you know about expansion and contraction, explain how Canadian pioneers made the metal rims fit tightly on the wooden wheels of their wagons.

Micro-organisms and Food

Imagine a totally different world. Think of yourself shrunk to the size of the period at the end of this sentence. At that size, you are still a giant compared with most of the living things that you will read about in this Unit—micro-organisms. You may not realize it, but there are many connections between the world you know and the largely unseen world of micro-organisms.

Most micro-organisms are not harmful to people. In fact we cannot live without some of them. Others are necessary to help us produce some foods such as bread, sauerkraut, yogurt, and cheese. But when food goes "bad," it is usually as a result of the action of micro-organisms. In this Unit you will look at various types of micro-organisms and find out how and why they affect food. You will also consider what can be done to keep food safe to eat.

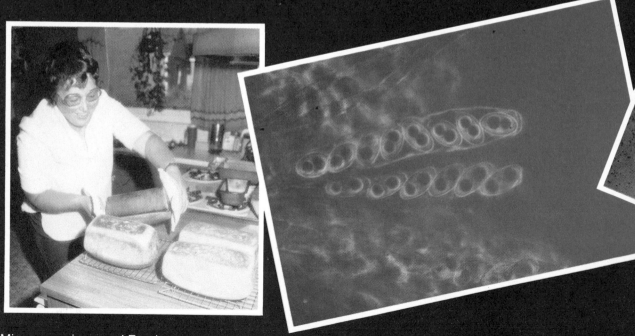

What Are Micro-organisms?

Imagine that you have just obtained a summer job. You are going to help take 40 children on a camping trip for one week. Your particular task is to take charge of the food for everyone on the trip. You send a questionnaire to all the parents of the children who are going on the trip. From the answers to the questionnaire you learn that the favourite food of most of the children is peanut butter sandwiches. Some of the parents have said that their children may not have food that contains preservatives.

You know that some bread will go mouldy in less than a week if it is not kept in the refrigerator. How are you going to be able to serve peanut butter sandwiches throughout the week of camping if you do not have a refrigerator in which to keep the bread? What could you do to keep the bread fresh? What are preservatives?

Knowing about micro-organisms will help you to solve your problem. Restaurant owners, physicians, scientists, bakers, and people who make the cheeses you like to eat all need to know about micro-organisms in order to do their jobs.

How Micro-organisms Were Discovered

The first person actually to see and describe micro-organisms was the man who invented the microscope in 1674. His name was Anton van Leeuwenhoek. He was a shopkeeper in Holland, and like other shopkeepers at the time, van Leeuwenhoek used a magnifying glass to inspect the quality of the cloth he sold. He became skilled at making high-quality glass magnifying lenses to use in his work.

Van Leeuwenhoek combined several of his small lenses to make the first microscope. With this microscope he obtained clear images of very small objects, magnified as much as 300 times or more. He discovered **micro-organisms** in samples of water and wrote about his discovery to the Royal Society of London, an organization of scientists. His letters to the Society created a sensation. Some scientists, however, found it difficult to accept his claim.

For many years after van Leewenhoek's amazing discovery, scientists peered down microscopes and observed more and more different types of micro-organisms. People came to accept the fact that we are surrounded by great numbers of tiny living things that we don't normally see. Eventually it became possible to recognize and classify these micro-organisms into five main groups, based on their characteristics. These groups are **protozoa, algae, bacteria, fungi,** and **viruses**. As you will see later in this Unit, many of these micro-organisms play an important part in our lives.

After van Leeuwenhoek had observed bacteria taken from the mouth, he made these drawings of what he had seen with his microscope.

Anton van Leeuwenhoek (1632-1723) was the first person to observe bacteria through his hand-made microscope.

How Small Are Micro-organisms?

Most people have never seen living micro-organisms in detail because they would need a microscope to do so. Micro-organisms are so much smaller than other living things that it is difficult to imagine their size. Even more surprising is the fact that there is a wide range of sizes among micro-organisms themselves! The next Activity will help you get acquainted with these microscopic forms of life.

Walking across the Head of a Pin

To get some idea of just how small micro-organisms are, try to estimate how many of them might fit across the head of a pin. To make this estimate easier, begin with a scale model of the head of a pin blown up to many times its real size.

Materials

masking tape
paper
scissors
tape measure
ruler
pencil
calculator

Procedure

1. Stick a piece of masking tape on the floor at one side of the room. Measure along the floor a length of 10 m and stick a second piece of masking tape on the floor. (If there is not enough room in your classroom to do this, you may have to use a larger room or the corridor.) The 10 m represents the diameter of a pinhead. Since the head of a typical pin is only 1 mm across, your model is 10 000 times bigger.

2. On a sheet of paper, draw a circle with a diameter of 2.5 cm. Cut out the circle. This represents the size of a typical protozoan on the model you used to represent the head of a pin.

3. Draw a rectangle 1 cm × 3 cm and cut it out. This rectangle represents a typical bacterium.

4. Draw a square 2 mm × 2 mm. Carefully cut it out. This tiny piece of paper represents the size of a virus on the model you used to represent the pin head. (Some viruses are even smaller—too small for you to be able to cut them out when using a model this size!)

5. When everybody in the class has cut out three different "micro-organisms," lay all the pieces of paper in a line on the floor. The line should run between the two strips of tape marking the edges of the pin head.

6. After you have answered the questions below, pick up all the pieces of paper and dispose of them properly.

Finding Out

1. How many "micro-organisms" are there walking across your model of the pin head?
2. Use your calculator to find out what percentage of the diameter of the pinhead these micro-organisms would cover if laid end to end. The following example will help you.

   If there are 30 students in your class,
   • the "protozoa" will have a total length of
   30×2.5 cm = 75 cm;
   • the "bacteria" will have a total length of
   30×3 cm = 90 cm;
   • the "viruses" will have a total length of
   30×0.2 cm = 6 cm.
   Therefore the total length of all the micro-organisms will be
   75 + 90 + 6 = 171 cm.
   The diameter of the pin head is 10 m or 1000 cm.
   $\frac{171}{1000} \times 100 = 17.1$
   Therefore, the micro-organisms will cover 17.1% of the diameter of the pin head.

3. From your answer to question 2, calculate how many of these micro-organisms could fit in a line across a pin head, if they were laid end to end.

Size Is Relative

In Unit One you found that organisms vary greatly in size. For example, the largest mammal is the blue whale, which can grow up to 3330 cm long. That is just over one-third the length of a football field. The smallest mammal is a kind of shrew, only 6 cm long, even including its tail. That is about the length of your little finger.

Although all micro-organisms are much smaller than the head of a pin, they vary greatly in size. The smallest micro-organisms are viruses. Some of them are about 500 times smaller than some of the larger micro-organisms—the protozoa. How does this difference in size compare with the difference between a whale and a shrew?

Types of Micro-organisms: a Guide to the Very Tiny

Can you tell the difference between a frog and a fish? These two animals are very familiar to you, so it is easy to recognize their differences. But when you first come to study micro-organisms, they may all seem to look alike. This is partly because many of them consist of only a single cell. Some types of micro-organisms, however, are much smaller than others. As you become more familiar with their appearance, you will begin to see other differences. To a **microbiologist**—a scientist who studies micro-organisms—one type of micro-organism is as different from another as a frog is from a fish. To help you sort them out, look at the pictures of the five main groups of micro-organisms shown on the next four pages and read about some of their characteristics.

Bacteria

Characteristics of bacteria

- All bacteria are single-celled.
- The cells of bacteria are very simple, and there are few structures inside them.
- There are three basic shapes of bacteria: spherical, rod-shaped, and spiral.
- Some bacteria can move by waving whip-like structures called **flagella**.
- Some bacteria live together in groups or clumps. There may be as few as four bacteria in a clump or as many as several hundred. When bacteria grow on a solid surface, they grow as a **colony**. Other bacteria live together in chains of cells.
- Bacteria often produce a slimy coating that protects them and holds them together.

Interesting facts about bacteria

- Some of the earliest forms of life on Earth probably looked like bacteria. In fact, fossils resembling bacteria have been found in rocks that scientists have dated as over 3000 million years old.
- The green scum on ponds consists of masses of blue-green bacteria.
- Many types of food poisoning and some diseases are caused by bacteria.

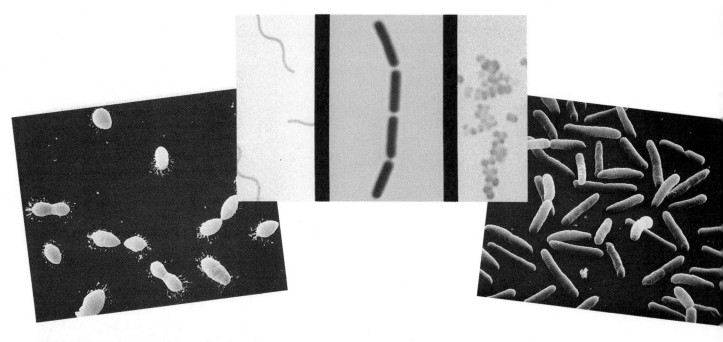

Viruses

Characteristics of viruses
- Viruses are many times smaller than bacteria and are even simpler in their structure.
- Viruses have many shapes: they may look like rods, spheres, cubes, or tadpoles, or have a number of other shapes.
- Viruses cannot move, reproduce, feed, or grow outside the cell of another living organism.

Interesting facts about viruses
- Viruses are so small they cannot even be seen with an ordinary microscope. Scientists have to use an electron microscope to study them.
- Most scientists do not classify viruses as living things. They are strange forms that exist somewhere between the living and the non-living world.

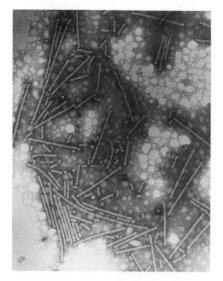

The tobacco mosaic virus was the first virus to be discovered. This photograph of the virus was taken through an electron microscope, which is many times more powerful than a light microscope.

Algae

Characteristics of algae
- Microscopic algae are single-celled.
- All algae contain several structures in their cells.
- There are many different shapes of algae.
- Most are coloured green, red, or brown.
- Some algae can move by waving one or two long flagella; some cannot move.
- Some algae live in colonies.

Interesting facts about algae
- The most common forms of algae are called **diatoms**. These microscopic algae often form a thick brown layer on rocks in the bottom of a river or stream. Many of these organisms have beautiful shapes and look like tiny crystals.
- When diatoms die, their hard cell walls sink to the floor of the ocean or lake. Over millions of years, countless numbers of these skeletons have built up deep layers of fine material called diatomaceous earth. This earth is used in many products, including polishing and scouring powders. The next time you clean the bathtub, you may be using micro-organisms to help you scrub!

Some typical diatoms. Diatoms are one of the most common living things on Earth.

Protozoa

Characteristics of protozoa
- Protozoa are single-celled and larger than bacteria.
- Most protozoa have several distinct structures inside their cells.
- They have a variety of shapes, and some, like the amoeba, can change their shape.
- Most protozoa can move. Some use whip-like flagella. Some use tiny, hair-like cilia.
- Some protozoa live together in colonies.

Interesting facts about protozoa
- A few species of protozoa can grow to be several millimetres in length and can be seen without a microscope.
- Some protozoa cause serious diseases, including malaria and dysentery.

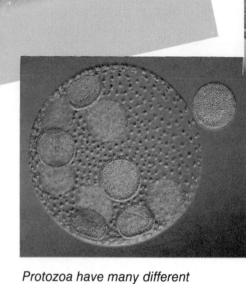

Structures in the cells of protozoa can be seen through the microscope as dark spots.

Protozoa have many different shapes and can move in different ways.

Fungi

Characteristics of fungi
- Some fungi, such as yeast, are single-celled. Others, like mushrooms, are **multicellular**; that is, they have many cells.
- Most fungi consists of microscopic threads or filaments, called **hyphae**, which look like a fuzzy mass of fine hairs
- Fungi have a variety of shapes, sizes, and colours.
- The majority of fungi cannot move.

Interesting facts about fungi
- Some fungi, such as bread mould and potato blight, spoil human food. Other fungi, such as yeast, help us make food.
- Fungi that you may be familiar with include those that cause mildew, apple scab, gill rot on fish, and athlete's foot.

Mushrooms are among the largest fungi.

These microscopic yeast cells are magnified many times.

Looking at Micro-organisms

Problem

What characteristics help you to distinguish various micro-organisms?

Materials

various samples showing micro-organisms
microscope
hand lens

Procedure

1. At labelled stations around the room are examples of the main groups of micro-organisms described earlier. Carefully examine the samples provided by your teacher. They may be in the form of photographs, microscope slides, or drawings. Use a microscope or hand lens when necessary.

2. Sketch each sample and label your drawing. Note the colour, shape, and size of each organism. When possible, record where it was found.

3. Refer to the previous pages for illustrations and information about the different groups of micro-organisms. You may also find it helpful to compare various samples with each other.

Finding Out

1. (a) Draw a table with five columns. Label the columns as follows: Bacteria, Viruses, Protozoa, Algae, Fungi.
   (b) Write in each column the characteristics that you observed in samples from that group (e.g., size, shape, colour, appearance, where found).

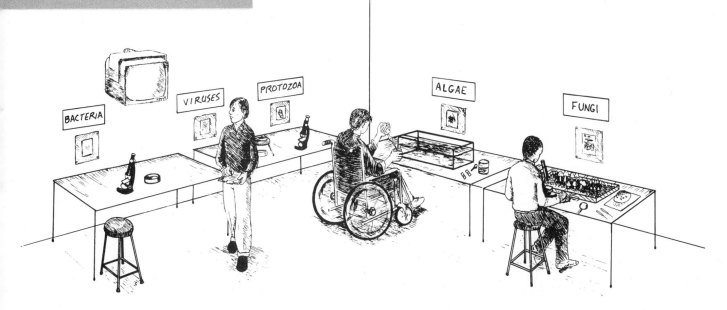

Micro-organisms Everywhere

Where Do Micro-organisms Live?

Wherever you go on Earth, you will never be alone. Micro-organisms will be with you. In the oceans, in ponds, in deserts, in forests, and in the soil, micro-organisms are found just about everywhere. There are even micro-organisms living in environments that would kill other living things. For example, some blue-green bacteria live in hot springs, like those in Yellowstone National Park, where the water temperature is over 75°C. Other bacteria live at below-freezing temperatures in permanent ice fields on mountain tops or at the North and South Poles.

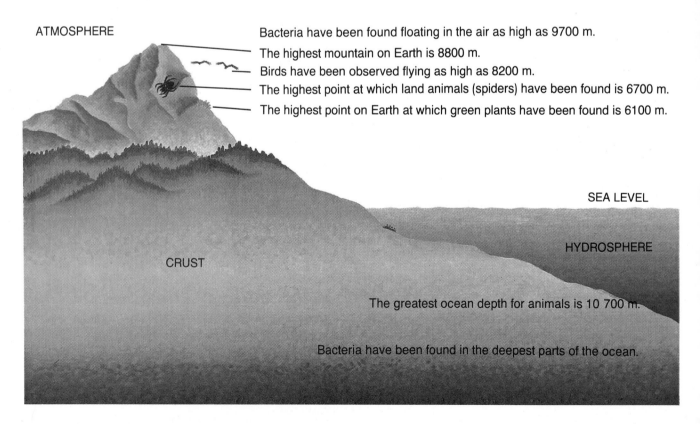

ATMOSPHERE

Bacteria have been found floating in the air as high as 9700 m.

The highest mountain on Earth is 8800 m.

Birds have been observed flying as high as 8200 m.

The highest point at which land animals (spiders) have been found is 6700 m.

The highest point on Earth at which green plants have been found is 6100 m.

SEA LEVEL

HYDROSPHERE

CRUST

The greatest ocean depth for animals is 10 700 m.

Bacteria have been found in the deepest parts of the ocean.

A few types of bacteria can survive extreme conditions by producing a resting stage called a **spore**. In this form, a bacterium is inactive and is protected by a thick cell wall. The spore remains at rest until suitable conditions return. Then the bacterium takes up its normal activities again. Bacterial spores can survive extreme dryness and will live even in boiling water or at temperatures as low as –250°C.

Bacterial spores have even been found on the ocean floor. Scientists estimated they were thousands of years old. Yet when provided with warmth, moisture, and food, the bacteria emerged from the spores, still alive! Some scientists think that life forms such as bacteria might travel as spores from planet to planet through outer space!

Many micro-organisms live on and in the bodies of animals and plants. For example, bacteria are common all over your skin, and especially around your mouth, gums, and teeth. They are also found in your digestive tract. Should you worry about this? Well, it's a good idea, of course, to wash regularly and brush your teeth and gums. Doing so will hold down the number of bacteria on the surface of your body and in your mouth. But the vast majority of micro-organisms are harmless. Some are even helpful to you. They break down certain types of food that you could not otherwise digest.

You will never be entirely alone — everywhere you go, there will always be micro-organisms, even in the desert.

Studying Micro-organisms

One way that scientists collect micro-organisms for study is to grow a **culture**, or colony, of micro-organisms. The culture is grown in a dish that contains food for the micro-organisms. Liquid food is first heated to kill any existing micro-organisms, then mixed with **agar**, a substance that solidifies at room temperature. The mixture is poured into dishes and cooled. The result is a dish of jelly-like substance that contains food but is **sterile**—that is, free from micro-organisms. A sample of the micro-organisms that the scientist wants to study is added to the dish, and the dish is covered to keep out any other micro-organisms. The Activity that follows gives you an opportunity to grow your own cultures.

The Micro-organisms Around You

Problem

What micro-organisms are found in everyday places and situations?

Materials

4 covered petri dishes containing sterile nutrient agar
sterile cotton swabs
coin
transparent tape
grease pencil
hand lens

CAUTION: While it is true that most micro-organisms are harmless, some are very dangerous. When scientists work with micro-organisms, they treat them all as dangerous until they have proved otherwise. Once you start growing cultures of micro-organisms in the petri dishes, you should treat all the dishes as though the cultures were harmful. Never remove the top once the dish is covered.

Procedure

1. Draw a sterile cotton swab over a surface where you think there may be micro-organisms, such as your desk, a window ledge, or a locker.
2. Raise the lid of a petri dish slightly, and lightly draw the swab across the surface of the agar. Immediately replace the lid of the dish and seal it with tape. With a grease pencil, write on the bottom of the dish the date and the place from which you took the sample.
3. If there is an aquarium or pond at your school, use a sterile swab to transfer some water to a second petri dish. Or, drag a coin over the agar. Seal and label the dish as before.
4. Lightly touch the agar in the third dish with the tips of your fingers. Seal and label the dish.
5. Wash your hands thoroughly with soap and water for 30 s; then touch your fingertips to the agar in the fourth dish. Seal and label the dish.
6. Leave all dishes *upside down* in a warm spot and check them every day for the next three or four days. Turn the dishes over and make careful notes and sketches of the appearance of the cultures growing in each dish. Use a hand lens to observe them more closely (but DO NOT remove the lids from the dishes).
7. Follow your teacher's instructions for proper disposal of the dishes at the end of the experiment.

Finding Out

1. Under suitable headings in your notebook, describe the shape, size, texture, and colour of the micro-organisms grown in each dish.
2. (a) What type of organisms (for example, fungi, bacteria) are in each sample?
   (b) How do you know?

How Many Micro-organisms Are There?

3. From which place did the fastest-growing culture come?
4. (a) What differences, if any, were there between the cultures that grew in the two dishes you touched with your fingers?
(b) How would you explain any differences you observed in (a)?

Finding Out More

5. (a) How do you think micro-organisms came to be in the places you took your samples from?
(b) Could you make any of these places sterile (free from micro-organisms)? If so, how?

Whereas micro-organisms are almost unbelievably small, their numbers are almost unbelievably large. They are the most common living things on Earth. A spoonful of garden soil contains many millions of bacteria. The total number of bacteria in your mouth alone is greater than the number of all the people that have ever lived. In fact, there are so many micro-organisms living on most animals that they increase the animal's mass!

One reason why micro-organisms are usually found in large numbers is that they can reproduce very rapidly. In suitable conditions, bacteria reproduce by dividing in two. Under "ideal conditions," some bacteria can grow to full size and divide into two bacteria in only twenty minutes. Because of this rapid rate of doubling, it is possible for huge populations of bacteria to build up in a very short time. In practice, of course, "ideal conditions" do not exist. Poisonous substances, diseases, shortages of food, and an unsuitable environment are some of the conditions that keep populations of micro-organisms (and of larger organisms as well) from becoming enormous.

If a bacterium divided every hour, there would be two bacteria at the beginning of the second hour. By the third hour, there would be four; by the fourth hour, eight; and so on. This "doubling" factor can be applied to many different kinds of situations. The following Activity examines one such situation.

Streptomyces antibioticus (bacteria) multiplying on a petri dish. Read about this micro-organism on pages 272 and 273.

Escherichia coli bacteria growing on agar.

Activity 5-4

Double or Quit

Problem

Imagine it is summer. School is over and you are planning things you could do. You begin to think about earning some money... Your neighbour, Ms A, asks you to help with some household chores. You will work for two days, eight hours each day. Ms A will pay you $20 an hour! Your other neighbour, Mr. B, also wants you to help for the same period of time. But his offer is even stranger. He will pay you 1 cent for the first hour, 2 cents for the next, 4 cents for the third, 8 cents for the fourth, and so on. Every hour, your rate of pay will be doubled. How much would each neighbour pay you for the sixteen hours?

Materials

calculator
notebook
graph paper

Procedure

1. On a sheet of paper, list the numbers from 1 to 16 to represent the number of hours you will work.
2. Beside the list of numbers, make one column for Ms A and one column for Mr. B. Write down the amount you would be paid by each neighbour for each hour. (Use your calculator to help figure out how much Mr. B would pay you.)

Finding Out

1. (a) Which neighbour would pay you more at the end of the first day?
   (b) Which neighbour would pay you more at the end of the second day?
2. (a) Which neighbour would you rather work for?
   (b) Does this answer surprise you? Why?

Finding Out More

3. Use your list of numbers from the column of payments made by Mr. B to draw a graph of the payments made for the first eight hours. Put time (in hours) on the x-axis, and amount of money (in cents) on the y-axis.

(b) Starting with 1, and doubling each time, what is the number you got after doubling 16 times?
4. Your graph shows the pattern of growth from doubling. In ideal conditions bacteria increase by doubling. The graph shows how exposed food can be infected by many bacteria in a surprisingly short time.
   (a) Predict how many times a bacterium must double to increase to more than one million.
   (b) Use a calculator to find out if your prediction in (a) is correct.

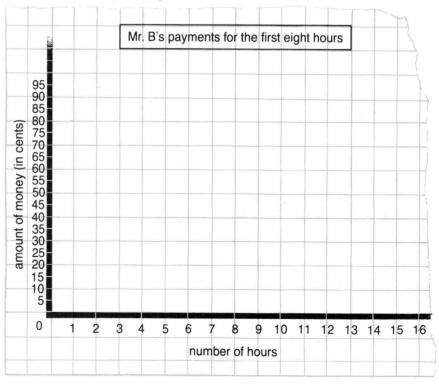

Mr. B's payments for the first eight hours

amount of money (in cents)

number of hours

Checkpoint

1. Match the micro-organisms in List A with the descriptions in List B.

LIST A	LIST B
1. algae	(a) one of the simplest, oldest, and most common types of organism.
2. fungi	(b) the smallest micro-organisms, considered by some scientists to be non-living.
3. bacteria	(c) vary from single-celled to multicellular; most have thread-like filaments.
4. viruses	(d) these micro-organisms include many different shapes and forms.
5. protozoa	(e) most are green, red, or brown.

2. (a) Why did people not know about micro-organisms 400 years ago?
(b) How might people at that time have guessed that micro-organisms were present?

3. When Susan told her younger brother she was studying micro-organisms at school, he replied, "Why study them? They're all so small and they're all the same." Explain why you agree or disagree with Susan's brother.

4. "Roberta Recycler flew down from the mountain, across the desert, and down to the edge of a large lake where people had been camping. Bits of orange peel were scattered around a picnic table. With a sigh, Roberta dug a hole, put in the waste, and covered it with soil. She washed her hands and smiled at a passing duck."

In this story, eight different places are mentioned where micro-organisms are commonly found. Write them down. Name two other places where micro-organisms are found.

5. (a) Give two reasons why micro-organisms are usually found in very large numbers.
(b) Name one possible result of rapid growth in a population of micro-organisms.

When algae, such as these pyrrophyta, reproduce rapidly, they can clog waterways and cause other problems such as "red tide," which gets its name from the colour of the algae.

Micro-organisms and the Environment

What would the world be like if there were no micro-organisms? Would we notice any change? You might think we would be better off because there would be no more rotting food. Soon, however, there would be serious problems. As you will learn in this Topic, micro-organisms are essential to many important biological processes. Without micro-organisms, in fact, none of the other forms of life could survive for very long.

Producers, Consumers, and Decomposers

All living things on Earth can be classified into three groups according to the way they obtain their food.

- **Producers** are organisms that produce their own food. All green plants and some micro-organisms produce their own food by the process of photosynthesis.
- **Consumers** are organisms that obtain their food by eating other organisms. All animals and some micro-organisms are consumers.
- **Decomposers** are organisms that obtain their food by breaking down wastes and dead material. Many micro-organisms are decomposers.

In the following Activity you will discover how the three groups of organisms interact in the environment.

Pass the Food, Please

Carefully study the following story; then answer the questions that follow.

After finishing a picnic lunch one day, Michael took one last bite of his apple and tossed it onto the grass. He thought no more about it.

What was waste for Michael was a feast for hundreds of thousands of micro-organisms. Mould and bacteria landed and reproduced on the apple. They started to break it down into the chemicals of which it was made. Gradually, the apple started to decompose.

*After many weeks, few traces of the apple were left. Most of the chemicals from which the apple was made had entered the soil. Some of these chemicals, called **nutrients**, had been taken up by nearby grasses through their roots. The grass plants needed the nutrients in order to grow.*

Finding Out

1. List the producers, the consumers, and the decomposers in this story.
2. Name two places where you would expect to find a lot of decomposers.
3. If there were no decomposers, what would happen to
   (a) wastes and dead matter?
   (b) the amounts of nutrients in the soil?

Finding Out More

4. (a) Continue the story of the apple, explaining what happens to the nutrients after they pass from the grass plants.
   (b) What do you think is meant by the term "cycle of nutrients"?
5. (a) Why do you think some people make a pile of grass clippings in their garden?
   (b) Why do farmers sometimes put manure on their fields?

Extension

Your teacher will show you some leaves that were buried in trays of soil several weeks ago. Some of the leaves were buried in ordinary soil. Some were buried in soil that had been sterilized by heating it in a hot oven for half an hour. Both trays of soil were kept covered, warm, and moist. Study and sketch the leaves, then answer the following questions.

6. Is there any difference in the appearance of the two samples of leaves? Explain.
7. (a) Why was the soil kept damp?
   (b) Why was the soil kept covered?

Micro-organisms as Producers and Consumers

You now know that the feeding habits of some micro-organisms are important to you. Without decomposers, dead animals and plants and waste materials would soon pile up in huge quantities. The feeding habits of other micro-organisms also affect you. There is a direct connection, for example, between your sitting down to a plate of fish and chips and some microscopic organisms living in a far-away ocean.

All organisms are connected to others in the environment by their relationship in a **food chain** such as the one shown here. Each organism in the chain depends on the next organisms in the chain for its food. Humans are part of many food chains that also include micro-organisms. People eat fish. But there would be no fish to eat if it were not for the smaller organisms that make up the food of the fish.

All food chains begin with the producers. Many micro-organisms, such as blue-green bacteria and algae, produce their own food. Without these producers at the beginning of each food chain, none of the other organisms in the chain could exist.

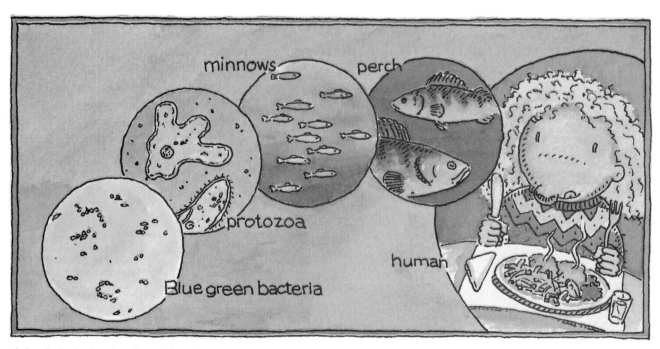

A food chain that begins with micro-organisms and ends with human beings.

Competing for Food

When our food begins to decay, it is because bacteria, fungi, and other micro-organisms are feeding on it. They are using *our* food as *their* food! Just like us, micro-organisms prefer certain foods. And just like us, they prefer certain conditions in which to live and eat.

Most micro-organisms grow best in environments that are warm, dark, and moist. Do the next Activity to test the growth of bread mould under these conditions.

Activity 5-6

Favourite Conditions

Problem

Does bread mould grow most quickly under moist, warm, and dark conditions?

Materials

slices of bread
petri dishes (or saucers and plastic wrap)
paper towels
hand lens

> **CAUTION:** Once you have started growing micro-organisms, never remove the lids of the petri dishes. Never touch, taste, or smell decaying food. Follow your teacher's instructions for disposing of your dishes at the end of the experiment.

Procedure

1. Have the class divide into three groups. Have each group choose one of the three sets of variables.
   • warm temperature or cool temperature;
   • dryness or moisture;
   • darkness or light.
2. (a) Plan an experiment to test how the two alternatives affect the growth of mould.
   (b) Decide on the other conditions you will need to control.
3. Set up your experiment and observe the bread samples daily over a period of a week. Record your observations and share them with the other groups in the class.

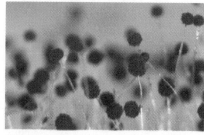

Examine your bread mould with a hand lens through the lid of the dish.

Finding Out

1. (a) Under what conditions did mould grow fast?
   (b) Under what conditions did mould grow slowly, or not at all?
2. Using your results, describe the conditions under which you would store bread.

Finding Out More

3. (a) Where do you keep bread in your home?
   (b) Is it always kept wrapped?
   (c) Do you think bread is stored in your home in a way that protects it from mould? Give reasons for your answer.

Extension

4. What else might affect the growth of mould on bread? For example, does mould grow faster on different types of bread, such as whole wheat, white, rye, or bread with no chemical preservatives? Does it make any difference if bread is covered or left exposed? You could answer these questions by conducting experiments.

From the human point of view, some micro-organisms have unusual tastes. For example, there are fungi that feed on cotton, paper, or leather. And there are bacteria that feed on wood, plastic, or oil. Several different micro-organisms feed on the food in the stomach of a cow. The millions of protozoa and bacteria in its stomach actually help the cow. They break down material in the grass that the cow cannot digest for itself. Without the micro-organisms in its stomach, the cow would starve.

Bacteria caused the Black Death, which struck Europe during the middle of the fourteenth century. This terrible plague wiped out a third of the European population. The person shown here is a physician. He did not know that the plague was caused by bacteria. He dressed in this way because he believed that each part of his costume would prevent the disease from entering his body. His cloak, hat, and open fingered-gloves were made of leather. The long beak of his mask was filled with aromatic spices, which he believed would purify the air passing through them. The eyes of his mask were made of glass. The physician held a wand in his bare fingers in order to take the patient's pulse.

Illnesses such as the plague are now mostly controlled by antibiotics (see pages 272 and 273). But even today, many millions of people still die every year of illnesses caused by micro-organisms.

Micro-organisms and Diseases

Many people think of bacteria as "germs" and believe they are all bad and cause disease. This is not correct. Many bacteria are probably harmless. It is true, however, that many diseases of animals and plants *are* caused by various kinds of micro-organisms. These organisms commonly cause disease symptoms by producing poisons, or **toxins**, that damage tissues or organs in the body.

Table 1 *Some Diseases Caused by Micro-organisms*

VIRUS	BACTERIA	PROTOZOA	FUNGI
common cold	bubonic plague	malaria	athlete's foot
measles	anthrax	amoebic dysentery	ergotism
AIDS	boils	sleeping sickness	ringworm
polio	strep throat		
rabies	pneumonia		
smallpox	tuberculosis		
warts	syphilis		
cold sores	cholera		
yellow fever	whooping cough		
chicken pox	tetanus		
influenza	scarlet fever		

Use your library to investigate
one of the diseases listed in
Table 1.
1. How does the micro-organism
   get to a person to cause the
   disease?
2. How can the disease be
   prevented?

Coughs and Sneezes Spread Diseases

Some diseases, such as the common cold, are spread through the air. The micro-organisms leave the body of an infected person when he or she sneezes or coughs. People nearby breathe in the micro-organisms and "catch" the disease. Other diseases are spread by touching food with hands that are not clean. How might the spread of these diseases be prevented?

Some micro-organisms that live in water cause diseases. When water contaminated with these micro-organisms is used for drinking or washing, people may become sick. Where the water supply is treated in ways that kill micro-organisms, there is far less disease from micro-organisms.

Activity 5-7

Lines to a Micro-organism

This poem was written by a Grade 7 student.

Do you know why mould
Is so bold
To grow even in the fridge so
    cold
And in the heat
On athlete's feet?

It is because
The mould has spores.
It needs no doors
But glides through the air
And lands on my sister's
    teddy bear.

The teddy bear
Had toast and butter in its
    hair.
It got soaking wet and
    steaming dry;
A perfect place for
    micro-organisms
To multiply!

Write a poem or story or draw a picture to show one or more of the roles micro-organisms play in the environment. You might portray micro-organisms as

- decomposers
- producers
- consumers
- agents of disease

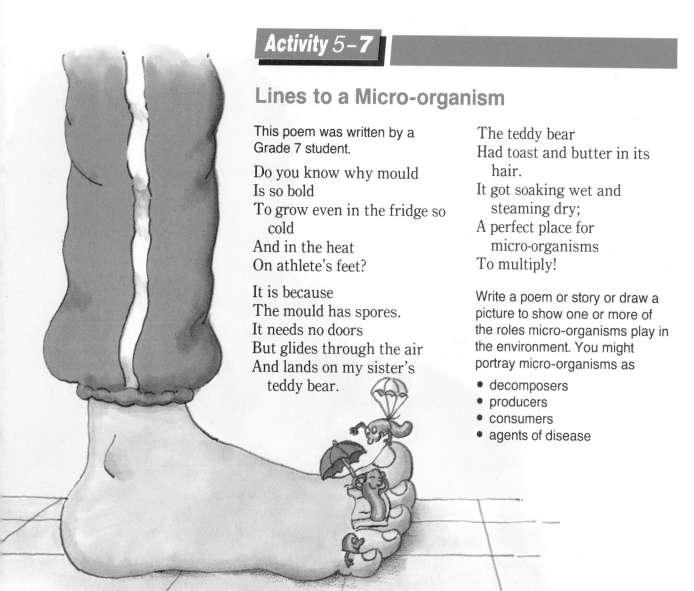

Was It an Accident, or Was It Good Science?

It's difficult to imagine that a mould like the one shown in the photograph can provide a substance that helps to control disease. But that is exactly what it does. Since it was first used about 50 years ago, an antibiotic from a mould similar to this has cured millions of people.

The antibiotic is penicillin—and it was discovered by accident.

Its discoverer, Alexander Fleming, was born in Scotland in 1881. After finishing his medical studies, he became interested in finding ways to kill bacteria that caused diseases in people. Unfortunately the chemicals so far discovered that killed the bacteria also killed healthy body cells.

During World War I Fleming carried out research on how wounds healed. He showed that the white blood cells helped to fight infections caused by bacteria in wounds.

In his later work Fleming showed that body fluids— tears, saliva, the mucus (liquid) inside the nose, and milk—could help to prevent bacteria from causing disease.

In the years following these discoveries, Fleming continued to look for a chemical that would be effective in combatting bacteria. In order to test various chemicals, he grew cultures of the bacteria *Staphylococcus*.

One day in 1928 as he was working in his laboratory, Fleming noticed that one of the culture plates had mould growing on it. Many people might have thrown the plate away, thinking it was useless. But Fleming noticed something startling. No bacteria were growing near the mould. Was the mould somehow killing the bacteria, he wondered?

Alexander Fleming had just discovered penicillin.

Immediately he started trying to separate the chemical that killed the bacteria from all the other

Mould growing on rye bread.

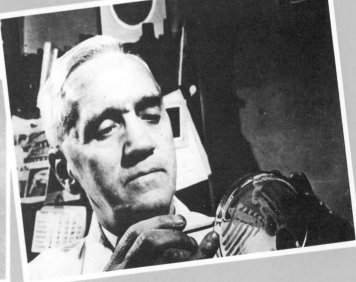

Alexander Fleming

chemicals in the mould. Unfortunately, he was not able to do so. Fleming was able, however, to test an extract of the mould in laboratory animals. He showed that the extract did not harm the animals, but did help to cure them of diseases. However, because he could not separate the bacteria-destroying chemical from the mould, he could not persuade other scientists to test penicillin on people.

World War II began in 1939. By 1940 many soldiers had received terrible wounds and it was vital to find ways to prevent these wounds from becoming infected. That year two chemists, Ernst Chain and Howard Florey, finally managed to separate the chemical penicillin from the other chemicals in the mould.

Penicillin was now a usable medicine with an outstanding ability to kill bacteria. It saved the lives of many soldiers during World War II. Today penicillin is used throughout the world to prevent bacterial infections of many kinds.

In 1945 Alexander Fleming, along with Ernst Chain and Howard Florey, received one of the highest awards in science—the Nobel Prize in Medicine and Physiology— for their work on penicillin.

Penicillin has not solved all our problems with diseases. It does not control all the diseases caused by bacteria. But since penicillin became well known, other antibiotics have been discovered that can control other bacterial diseases.

Penicillin and the other antibiotics act only against bacteria. They do not control diseases caused by viruses, such as cold viruses. The future challenge is to find substances that can control these viruses. Will a keen observer, like Fleming, notice something that no one else has noticed, and set out to investigate it?

Think About It

Was Fleming's discovery really accidental? What do you think?

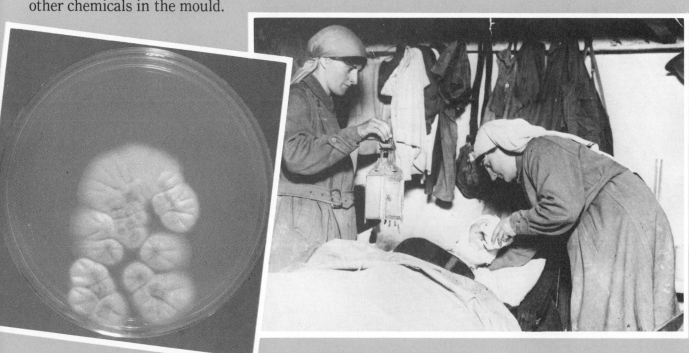

Penicillin culture growing in a petri dish.

Caring for a soldier wounded during World War I.

Micro-organisms and Food

Micro-organisms that Spoil Food

When micro-organisms decompose our wastes, we regard it as a very good thing. When they decompose our food, however, we regard that as a very bad thing. We say the food has gone bad or is spoiled, rotten, or sour. To the micro-organisms, however, kitchen wastes and your lunch are much the same. The challenge for human beings is to keep the micro-organisms that "spoil" our food off or out of it until we have completely finished with it.

Food that has been invaded by bacteria and fungi is not just unpleasant to see or smell. It can also be dangerous if it is eaten. Some of the micro-organisms that spoil food can make people very sick or even kill them. The result of eating food infected with some bacteria is **food poisoning**, or food-borne illness.

Spoiled food! yum!!

Types of Food Poisoning

There are three common types of food poisoning: "staph," botulism, and salmonellosis. They are caused in different ways, but all of them can be prevented by handling and preparing food carefully.

Write down two things in this picture that could lead to food poisoning. How could this risk be avoided?

The full name for "staph" is **staphyloccal food intoxication**. It is the most common type of food poisoning. It is caused by bacteria that normally live on the human skin and in the nose and throat. The bacteria get into food when people handle the food without washing their hands, or when they cough or sneeze over food. The bacteria grow and multiply rapidly in many foods, especially dairy products like cream. They produce toxins that cause nausea, diarrhea, and abdominal cramps a few hours after the infected food is eaten. The victims usually recover completely in a few days.

Botulism is caused by a type of bacteria that is common in the soil. The bacteria are not dangerous unless they are trapped with food under **anaerobic** conditions—that is, in a place where there is no oxygen. For example, people often can or bottle food at home. They heat the food and seal it from air (and oxygen) in bottles or jars. If they do this without proper care, the bacteria may produce a toxin that can kill. The botulism toxin is destroyed by high temperatures. For this reason, food should be thoroughly cooked before being sealed into jars or bottles. The containers should also be washed in very hot water. An extra safeguard is to cook preserved food again just before eating it.

Salmonellosis is caused by the salmonella bacteria themselves, not by a toxin. When these bacteria are swallowed with food, they start reproducing in the digestive system. In large numbers they cause nausea, diarrhea, and fever.

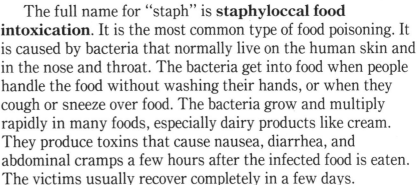

What is the greatest risk of food poisoning in this situation? How can it be avoided?

Come and watch this great hockey game!

What type of food poisoning might result from this situation. How can it be prevented?

Salmonella bacteria also live in farm animals like chickens or cows. People can get sick by eating the infected meat of these animals, or by eating infected eggs or dairy products such as milk. Many cases of salmonellosis have been traced to situations in which food was left standing in a warm place for some time. The bacteria thus had ideal conditions in which to reproduce in large numbers.

Food must be kept and handled in clean conditions. Otherwise it can become contaminated. The food industry has to make sure that food is handled safely before it gets to the customer. Sometimes, however, things go wrong. In the following Activity, you can investigate some of these problems.

Activity 5-8

Keeping It Clean

You work for a restaurant chain. Your job is to make sure that the food served in all the restaurants in the chain is free from contamination that might make your customers sick.

In one of your restaurants, there have been several complaints from customers during the last month. The customers reported upset stomachs, which they claimed were caused by the food served in the restaurant. Other, similar restaurants in the same city had not received any complaints. The restaurants all get their food from the same food suppliers. Therefore, the contamination must be taking place in the restaurant itself, not at the suppliers.

You have been asked to investigate the situation.

Here are some suggestions that you might want to consider as you carry out your investigation. Using any of these ideas, as well as ideas of your own, draw up a plan in writing for your investigation.

- List all the places in the restaurant where food is found.
- Beside each place, write how the food is handled in that place.
- Make a list of the people who work in the restaurant, and of the job each does.

- Draw up a list of rules that the people who handle the food must follow.
- Make a list of questions you might want to ask about how food is prepared and handled in the restaurant. As you make your list, think about other information you might need in order to draw up your plan of investigation.

Now complete your plan.

Micro-organisms that Produce Food

Over 3000 years ago, people observed that if certain soft fruits were mashed and left in a warm place they soon started to produce bubbles. If the fruit juice was left long enough under the right conditions, it eventually became wine. In Latin, this foaming was called *fermentare*, meaning "to cause to rise." From this, we get our word **fermentation**.

We now know that microscopic yeast cells cause the fruit juice to ferment. These single-celled fungi are carried through the air by the wind. When the yeast cells land on something sugary, such as the fruit juice, they start to feed on it. In doing so, they cause a chemical reaction that produces alcohol and carbon dioxide gas. It is this gas that causes the bubbles during fermentation.

Yeast is not the only micro-organism that has the ability to cause fermentation, and alcohol is not the only product of fermentation. There are many different kinds of fermentation, and many different kinds of micro-organisms that cause it. In each case, the micro-organisms bring about a change in the flavour of the food.

Do you like cheese? Or chocolate? These products, together with many other foods such as yogurt, bread, pickled vegetables, certain sausages, beer and wine, are all made with the help of micro-organisms.

Looking at Fermentation

Problem

Under what conditions will yeast cause fermentation?

Materials

test tube rack
4 test tubes
4 balloons
packet of dried yeast
sugar
10-mL graduated cylinder
distilled water or boiled tapwater
labels
set of metric spoons

Procedure

1. Blow the balloons up a few times to stretch them.
2. Prepare the test tubes as shown. Use about 10 mL water in each of (a), (b), and (d). Label the test tubes and stand them in a warm place for one or two days.
3. Copy Table 2 into your notebook.
4. Predict which combination or combinations of materials will produce fermentation.
5. Examine the tubes and record in your notebook any changes you observe.

Table 2

CONTENTS OF TUBE	BALLOON INFLATES	BALLOON DOES NOT INFLATE
(a) sugar and water		
(b) yeast and water		
(c) yeast and sugar		
(d) water, sugar, and yeast		

Finding Out

1. (a) Which tube or tubes produced a gas?
   (b) How do you know?
2. Assuming that the gas produced is carbon dioxide and that it is a product of fermentation, list the conditions that are needed for fermentation to occur.
3. Is this a controlled experiment? Explain. (Read *Skillbuilder Two* on controlled experiments.)

Finding Out More

4. (a) Use your observations to explain why bread dough containing yeast rises.
   (b) Look at a piece of bread with a hand lens. Why are there many small holes in the bread?
5. (a) Name one way in which fermentation is similar to decomposition.
   (b) Name one way in which these two processes are different.

(a) 2 mL sugar + warm water

(b) 2 mL yeast + warm water

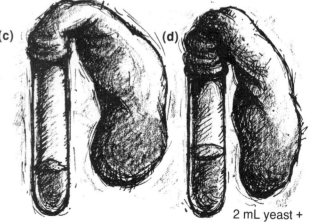

(c) 2 mL yeast + 2 mL sugar

(d) 2 mL yeast + 2 mL sugar + warm water

How Is It Made?

Did you ever wonder how some of the foods in your local supermarket are actually made? For example, how is milk turned into cheese or yogurt? Now is your chance to find out.

Cheese, many other dairy products, bread, and sauerkraut are examples of food made with the help of various micro-organisms. The questions on these pages are suggestions to help you begin finding out how these foods are produced. If you prefer, you can choose another type of food that is made by fermentation. You can find information about the food from

- your library
- Agriculture Canada
- your local supermarket
- companies that produce these foods

Your goal is to discover how micro-organisms are involved in the production of the food you choose. If possible, find out

- the type of micro-organism(s)
- its name
- the particular ingredients that it ferments
- the conditions under which fermentation occurs

Work in small groups and prepare a presentation of your findings to the class. You may want to include tables, pictures, or samples in your presentation.

Cheese

1. (a) What is the main ingredient in cheese?
   (b) How are bacteria used in cheese-making?

(c) Which cheeses are made by using mould?
(d) How long does it take to make cheese?

Other Dairy Products

2. (a) What micro-organisms are used to make buttermilk, sour cream, and yogurt?
   (b) What substance gives them all a slightly sour or acidic flavour?
   (c) What ingredients are listed on a container of yogurt?
   (d) How can you make yogurt at home?
   (e) What might you see if you looked at yogurt under a microscope?

Bread

3. (a) What micro-organisms are used in bread-making?
   (b) What conditions in a bakery encourage the growth of these micro-organisms?
   (c) How do the micro-organisms help to produce bread?
   (d) What products of fermentation are produced during bread-making?
   (e) What happens to these products during baking?

Sauerkraut

4. (a) What ingredients are in sauerkraut?
   (b) What products of fermentation give sauerkraut its flavour?
   (c) What is the difference between alcohol and vinegar?

Cheese being made in a dairy.   *Bread being made in a bakery.*

Checkpoint

1. Solve the puzzle in your notebook by supplying the correct word for each clue. When you have completed the puzzle, you will discover one of the subjects of this Unit.

(a) ☐ ■ ■ ■ ■ ■ ■ ■
(b) ■ ☐ ■ ■ ■ ■ ■ ■ ■ ■ ■ ■
(c) ☐ ■ ■ ■ ■ ■ ■ ■ ■
(d) ☐ ■ ■ ■ ■ ■
(e) ■ ☐ ■ ■ ■ ■ ■ ■ ■ ■ ■
(f) ■ ■ ☐ ■ ■
(g) ■ ■ ■ ■ ☐
(h) ■ ■ ■ ■   ■ ■ ☐ ■ ■

(a) Food poisoning caused by micro-organisms under anaerobic conditions
(b) A kind of food poisoning that is caused directly by a micro-organism, not by a toxin
(c) Organisms that obtain their food by eating other organisms
(d) The substances produced by micro-organisms that cause food poisoning
(e) A process used in preparing food that produces alcohol and a gas
(f) The smallest kind of micro-organism
(g) Micro-organisms that consist of thread-like hyphae
(h) A relationship in which a series of organisms depend on other organisms for their food

2. Early one summer, you notice a dead butterfly lying near a fence in the corner of the yard. A few weeks later, in the same spot, you see only a few faded patches of butterfly wings. How would you account for this change? What has probably happened to the rest of the butterfly?

3. (a) Draw a sketch to show how the following organisms might be connected in a food chain.
wolf
grass
deer
(b) Draw a sketch to show how the following organisms might be connected in a food chain.
bacteria
shrimp
protozoa
fish
otter

4. Which organisms in the sketch you have drawn for Question 2 are:
(a) producers?
(b) consumers?
(c) decomposers?

5. You are visiting a friend who is sick with a cold. During your visit your friend sneezes a lot, then picks some grapes from a bunch on the bedside table and hands them to you.
(a) Why should you be worried?
(b) Name two things you should do to avoid getting sick.

6. Write down three actions that could lead to food-borne illness. Explain why they would do so.

7. "A micro-organism landed on some food and began to digest it. The micro-organism multiplied. Some time later, Gordon Gourmet came by and ate the food. It was delicious."
(a) Explain the situation in this story. What sort of micro-organism might have been involved?
(b) What happened to the food?
(c) Name three foods that this story might be describing.

8. In a class studying businesses, Michaela wrote down these definitions.
• The *producer* is the company that makes the product.
• The *consumer* is the person who uses the product.
• A fast-*food chain* is a group of restaurants associated with each other that serve food.
Give the meaning each of these terms has in science.

Keeping Food Safe to Eat

Food is perishable. That is, it does not last forever and at some point it begins to decay. Decay is caused by micro-organisms. From the beginning of human history, people have looked for ways to prevent food from spoiling. Food can be protected:

- by keeping micro-organisms out of it;
- by killing or slowing the growth of micro-organisms that are already in it.

Examine these pictures. They show three methods used to preserve food. Each method discourages micro-organisms from growing in the food by depriving them of at least one condition that they need in order to grow. Which foods deprive micro-organisms of moisture? Which deprive them of warmth? Which deprive them of oxygen?

Cooling, Fermenting, and Drying

Packaging, canning, and freezing are fairly recent methods of preserving food. Many other methods we use for preserving food, however, have been used for hundreds or even thousands of years.

Long before ice boxes and refrigerators were invented, people knew that food kept at cold temperatures took longer to spoil than food that was warm. They stored food in cool places like cellars and basements, or packed it in snow or ice. In 1880, a cargo of meat being shipped from Australia to Britain in a cold container accidentally froze solid during the journey. Whereas some of the meat usually spoiled during such a long voyage, the freezing had kept all of it from spoiling. Thus, freezing came to be used for long-distance shipment and storage of food. Why does food kept in the cold take longer to spoil?

You have learned that fermentation is a way to make certain foods. It is also a method of preserving food. For example, cheese, which is made by fermenting milk, does not spoil as quickly as milk. Wine keeps much longer than the fruit juice from which it is fermented. Vegetables pickled in vinegar last longer than fresh vegetables. The reason is that the acid or alcohol produced by fermentation prevents the growth of most micro-organisms.

This fruit juice will start to ferment given the right conditions.

A cold storage area for cheeses.

Fish can be kept for a short time in crushed ice.

Without moisture few micro-organisms can survive. Traditionally many foods were preserved by drying them. Some fruits were dried in the sun. For example, dried grapes became raisins, and dried plums became prunes. Meat and fish were sometimes smoked, as well as dried, by hanging them in the smoke from a wood fire. Fish, especially cod, was commonly preserved by rubbing salt into it. How would salt help preserve it?

The next Activity allows you to see for yourself how effective some different ways of preservation are.

Fish and meat were commonly preserved by hanging them to dry, or by rubbing salt into them. How would salt help preserve them? Another technique was to hang fish and meat in smoke from a fire. How would this help preserve them?

Activity 5-11

Taking Out the Water

Problem

How can salt and sugar be used to dry food?

Materials

3 thick slices of fresh potato
3 jars with lids
salt
sugar
labels or grease pencil

Procedure

1. Place one piece of potato in each jar.
2. Completely cover one slice with salt, and a second with sugar. Do not put anything on the third slice. This slice is your control.
3. Put on the lids of all three jars and label them *sugar, salt, control.*
4. Record the size and appearance of each slice of potato at regular intervals over the next two or three days.

Finding Out

1. What happened to
   (a) the slice covered with salt?
   (b) the slice covered with sugar?
   (c) the slice left as a control?
   Explain any changes you observed.
2. What would happen to the micro-organisms on each slice? Why?
3. Why do you think you were asked to use jars with lids?

Finding Out More

4. (a) Make a list of foods you see in the supermarket or have at home that have been preserved by drying.
   (b) Beside each food, write the method used to dry it (for example, air drying, salting, smoking).

Keeping Food Safe to Eat   **283**

The jar invented by John Mason in the United States in 1858. After food was packed into the jars, the jars were sealed and put into boiling water. The heat killed most micro-organisms in the food, and the air-tight seal prevented other micro-organisms from getting in.

cross-section of the wall

lacquer on inside

tin plate

steel

structure of the can

sealing the edge

Modern cans are made of thin sheet steel sandwiched between tin or aluminum. There is a double seam at the top and bottom to make sure there are no gaps for air to enter. The cans are lined with materials that resist corrosion.

Cans and Chemicals

Methods of preserving such as salting, drying, and fermenting make it difficult or impossible for micro-organisms to grow in food. But these methods also change the taste of the food. How could you protect food from micro-organisms without altering the food itself? That is not such an easy task.

Most of the micro-organisms that are found in food get there after being carried through the air, or are passed on from the skin when food is handled. The longer food is left exposed, the greater the chance that it will be infected. Therefore, if food is to be stored for a long time, it must be shielded both from the air and from human hands.

Although for hundreds of years people have used containers to store food, these containers were not usually tightly sealed. The first mass-produced container that could easily be made air tight was the glass jar.

About the same time that jars were first used, people were experimenting with using metal to make food containers. In the early days there were problems with cans:

- cans could be dented, leaving small holes through which air—and micro-organisms—could enter;
- cans were sealed by melting the edges of the metal with solder, a method that sometimes left small gaps along the joint;
- the lead in the solder sometimes contaminated the food;
- cans could rust;
- the metal in cans could give food an odd taste;
- acidic foods like apples and plums might corrode the inside of the can.

Gradually, with the development of better materials and methods of sealing the cans, most of these problems were overcome. Almost every type of fruit, vegetable, meat, and fish can now be canned successfully.

The addition of chemicals to food to prevent the growth of micro-organisms is not a new method of preservation. It began long ago when people used vinegar, salt, sugar, alcohol, and spices to preserve food. These substances all occur naturally.

Today there is a wide range of **chemical additives** that are added to food to keep micro-organisms out of it. Chemical additives are chemicals made by people. People have allergic reactions to certain of these chemical additives more often than to naturally occurring preservatives such as salt and vinegar. Food companies are constantly looking for new and better chemicals to use, but it may take several years of testing and

In 1984 and 1986 a team of researchers from the University of Alberta made a fascinating discovery in the Canadian Arctic. On Beechey Island in the Northwest Territories, they were able to dig up and examine the graves of three British sailors who had been buried in 1846. It is now thought that one of the causes of the illness and deaths of the sailors was lead poisoning from the cans of food carried in the ship. These sailors were part of an expedition headed by Sir John Franklin. Like many others before him, Franklin had tried and failed to find a water route across the Arctic.

use before the possible harmful side effects of a new chemical are known.

In the next Activity, you will examine the arguments for and against a specific case of using chemicals to fight micro-organisms on food.

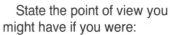

Activity 5–12

Scabby Apples

Apple scab is a fungus that causes a scab-like brown growth on the surface of apples. It is not known to cause any harm to human health. Most people in Canada, however, will not buy scabby apples even at a low price, because they do not like the appearance of scabby apples. A substance called fungicide can be sprayed onto the apples to kill the fungus. The use of the fungicide adds to the cost of the apples. The fungicide may also be harmful to human health, especially to people with allergies.

Should a fungicide be used on apples or should it not? To explore this issue, discuss the points below and decide on the action you would take to try to find a fair solution.

State the point of view you might have if you were:
- an apple grower
- a grocer selling apples
- a person who is allergic to fungicides
- a trucker who drives a truck taking apples to market
- a manufacturer of the fungicide
- your grandmother, who loves apples
- yourself

Extension

1. Imagine that you are a scientific expert on apple growth and fungicides. You are asked to give your advice on this issue. List the information you feel you would need before you could give your advice.
2. Using what you know about the use of fungicides on apples, and the effect of fungicides on human health, write a report giving your advice on what to do.
3. Imagine that you are the Minister of Agriculture for the province. Consider all the opinions of the people who are involved with apples, and the expert scientific advice you have received. Evaluate the alternative courses of action that have been proposed. What further questions might you have?

Even though the fungus that causes apple scab is not known to harm people, most people prefer apples that look like those on the right.

Handling Food

You discovered in Activity 5–3 that there are micro-organisms on your skin. Think about everyday habits of hygiene. Why should you wash your hands before touching food? Why do people wrap or cover food? You can use your answers to these questions in the next Activity.

Activity 5–13

Micro-organisms Are Not Invited

Problem

How can you keep micro-organisms from spoiling food?

It is a warm summer day, and some of your friends are coming over late in the afternoon for a meal. Because of your schedule, you want to prepare as much of the meal as you can in the morning. Here is the menu:

chicken noodle soup (from a can)
cheese sandwiches
potato chips
fresh fruit salad
whipped cream
soda pop

Procedure

1. Examine the menu and list all the ingredients in your meal.
2. Draw a table like Table 3 and list, in order, all the steps you plan to take to prepare the meal. Consider which foods you will prepare first, and which can be left until later. How will you protect the prepared food from micro-organisms before your guests arrive? For each step, provide a reason for your actions. The first step is given as an example.

Table 3

MORNING		
STEP	REASON	
1. wash hands	removes micro-organisms from skin	
2.		
AFTERNOON		
STEP	REASON	
1.		
2.		

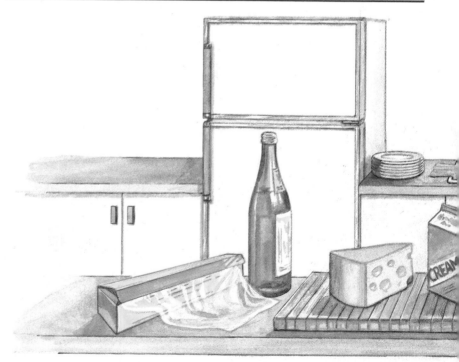

From the Plant to the Plate: Keeping Food Good to Eat

The food on your dinner plate may have been grown in your own province. It may have come from some other part of Canada. Or it may have travelled halfway around the world to reach your table. Unless you grow all your own food and eat it fresh, you depend on many people to make sure that your food is tasty and safe to eat.

Think about a can of stewed tomatoes, such as you might use to make spaghetti sauce. The tomatoes were grown on a farm. Once they were ripe, they were picked, packed, and shipped to a factory. There, they were washed and cooked. Spices and other things were added to the cooked tomatoes, and they were sealed into a can. The cans were packed in boxes and stored in a warehouse before being trucked to the store where you bought them. On their journey from the farm to your plate, the tomatoes were handled by many different people and different industries. They included the farmer, the food processing factory, the warehouse and trucking companies, and the storekeeper. The people at each of these various stages have the same problem you do in your kitchen. That is, how can the food be kept from spoiling?

Finding Out

1. Which of your steps involved:
   (a) preventing micro-organisms from getting into the food?
   (b) killing micro-organisms in the food?
   (c) slowing the growth of micro-organisms in the food?
2. How might the package in which food comes from the store help you reduce the chance of contamination by micro-organisms?
3. Name one other method you could use to protect food from micro-organisms.

Probing

Two modern developments in methods of preserving foods are **freeze-drying** and **irradiation**. Choose one of these methods to investigate.
(a) How does the method protect the food?
(b) What process does the method use?
(c) Can it be used for all foods or only some?
(d) Can you find examples of food preserved by this method in your local supermarket?
(e) What are the advantages and disadvantages of the method?

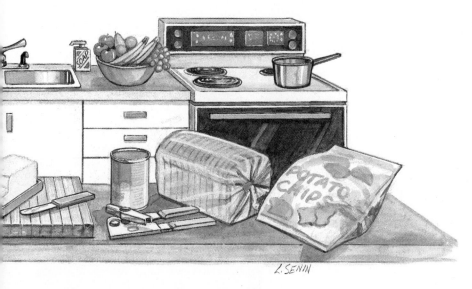

L. SENIN

Will That Be Smoked or Frozen?

Problem

How can you decide which is the best method of preserving and packaging food?

You are the owner of a food processing plant. In Table 4 are listed some of the foods you have in stock and some of the types of customers you sell food to.

Each type of food can be preserved in several different ways. For example, it could be dried, frozen, canned, smoked, or pickled. How can you choose the best method to meet the needs of each customer?

Copy the table into your notebook and complete it. The following questions may help you to choose the best method in each case.

- Does the method affect the taste of the food?
- Does it affect the mass of the food and, therefore, the cost of shipping?
- Does the method use expensive packaging or costly ways of storing the food?
- What other questions might you want to consider in making your choices?

Finding Out

Where two or more methods were equally suitable, what factor did you consider most important in making your final choice?

Table 4

FOOD	CUSTOMER	METHOD OF PRESERVING	REASON
vegetable soup	campers, hikers, canoeists		
vegetable soup	someone who wants a quick, cheap, easily prepared meal		
baby cucumbers	someone who wants them to last for one year		
fish	someone who wants them to last for one year		
fish	owner of a fish and chip restaurant		
strawberries	someone who wants to make jelly or cake		

In the mid-1800s, some French wine makers had a problem. Their wine was turning vinegary and sour. They asked a local professor of chemistry, Louis Pasteur, to find out what caused the souring. Using a microscope, Pasteur observed both ordinary wine and sour wine. He noticed that the ordinary wine contained many yeast cells; the sour wine also contained large numbers of a smaller organism. By further observation and by experiment, Pasteur discovered that the smaller organisms were bacteria. These had accidentally got into some batches of wine. The bacteria produced an acid that made the wine taste sour.

Pasteur is considered to be the founder of the science of microbiology, the study of micro-organisms. He made a great number of discoveries and inventions.

Louis Pasteur working beside a microscope in his laboratory.

Setting the Standard

At one time the people who sold food could prepare and handle it in any way they liked. There were no laws setting standards for cleanliness or for the quality of the food. Customers never even knew for sure what they might be getting! For example, tea was once expensive and in short supply, because it had to be brought by slow sailing ships from India. To save money and to increase their supply, some merchants bought used tea leaves from servants in large households. They mixed these used leaves with their fresh ones. To make the used leaves appear fresh, they dried them and coloured them black with lead. We now know that lead, unfortunately, is very poisonous.

By the beginning of this century, scientific studies had made most people aware of the dangers of contaminated food. Governments began to pass laws to control businesses that handled food. Since then, governments have carried out regular inspections of grocery stores, restaurants, and factories to ensure that the standards are being met. The purpose of these standards is to protect people from buying and eating harmful substances. But it isn't always easy for governments to decide exactly what practices and what ingredients are harmful.

What Is Safe?

You know that you can protect food from micro-organisms by keeping it in the refrigerator or by heating it on the stove. Knowing this, however, is not quite enough. How well these methods work also depends partly on the type of micro-organisms and on the type of food.

As you have learned, some micro-organisms can live in hotter or colder conditions than others. At –18°C, for example, some types of yeast cells are killed, but many bacteria and moulds simply become inactive. They will become active again at warmer temperatures. At the other extreme, some micro-organisms are killed at temperatures of 60°C, while others can survive boiling. For these reasons, recommendations on how to treat and store food are helpful. Do you recognize any of these names from Topic Four?

Table 5 *Heat Sterilization Table*

MICRO-ORGANISM	TIME TO STERILIZE (min)	TEMPERATURE (°C)
Salmonella typhosa	4	60
Lactobacillus bulgaricus	30	60
Clostridium botulinum	100–300	121 (steam)

Just as you have your favourite foods, so do micro-organisms. Some foods, therefore, start to decompose much more quickly than others kept under the same conditions. Governments publish tables that recommend how long different foods can be safely stored. Table 6 shows how long certain fresh foods can be kept safely in a refrigerator; some of them could be kept for much longer in a freezer.

Table 6 *Recommended Maximum Time to Store Different Foods in a Refrigerator*

cabbage	4 weeks		corn	2 days
carrots	2 weeks		ground beef	1-2 days
tomatoes	1 week		chicken	4 days
celery	5 days		oranges	2 weeks

These guidelines are important, because you cannot always be sure from looking at a piece of food whether it is contaminated or not. Which of the foods in the table would you eat after two weeks?

INGREDIENTS: WHOLE WHEAT, LIQUID SUGAR, LIQUID INVERT SUGAR, MALT SYRUP, SALT, NIACINAMIDE (21.4 mg/100 g), REDUCED IRON (IRON 14.3 mg/100 g), THIAMINE MONONITRATE (VITAMIN B₁, 2.2 mg/100 g), RIBOFLAVIN (VITAMIN B₂, 3.6 mg/100 g). BUTYLATED HYDROXYTOLUENE IS ADDED TO THE PACKAGE MATERIAL TO HELP MAINTAIN FRESHNESS.

TYPICAL NUTRIENT VALUE
PER SERVING

Food energy	108 Calories
Protein	2.7 g
Fat	0.2 g
Total carbohydrates	23.7 g
Sucrose & other sugars	5.0 g
Starch & other carbohydrates	16.9 g
Dietary fibre	1.8 g

VITAMINS AND MINERALS
PER SERVING

Vitamin B₁, Thiamine	0.6 mg
Vitamin B₂, Riboflavin	1.0 mg
Vitamin, Niacinamide	6.0 mg
Iron	4.0 mg

Serving size = ⅝ cup = 1 oz = 28 g;
Protein 9.5 g/100 g; Fat 0.8 g/100 g;
Carbohydrates 84.8 g/100 g;
Food energy 384 calories/100 g.

Probing

1. Examine some food labels from home or the supermarket. For each label, record
   (a) the main food ingredients
   (b) any other contents of the package (additives)
2. (a) Find examples of packaged food in your local stores or at home that do not have additives. (They may be described on the label as "pure" or "no preservatives added.")
   (b) What kind of products are they?
   (c) What difference does the lack of additives make to
   • how fresh the food stays?
   • the way the food is stored?
   • the taste or appearance of the food?

Taking Control

Standards for quality are set by governments. Government regulations describe the conditions under which food must be prepared and stored. In many cases there is a date on the package that recommends the time by which the food should be eaten.

Governments also control the use of particular ingredients. Some additives, for example, can be used only in quantities that have been shown by experiment to be safe. Check the labels on cans or packages at home or in the stores.

Sometimes **contaminants**, such as pesticides, accidentally get into food while it is being grown or processed and cannot be removed. A contaminant is any material that is unintentionally included in the product. Governments have to decide what quantities of these substances are acceptable. They do so on the basis of the knowledge available at the time.

The contents of all packaged foods, including additives, must be listed on the package. This list allows customers to know exactly what they are buying. Knowing what the food contains is especially important for people who have allergies or who must not eat large amounts of certain ingredients, such as salt or sugar.

A Last Word

At many different stages foods can be contaminated by contact with micro-organisms: bacteria in soil; micro-organisms carried on human hands, on insects, rodents, or other pests; micro-organisms on unclean surfaces, in water, and in the air. Because of one form of spoilage or another, much food that is grown never reaches the stores.

By the time the food is in the store ready for you to buy, government standards, controls, and inspections have prevented most serious problems. But food standards don't end there. Now your responsibility as the consumer begins. Do you know how to choose the best-quality food products at the store? Do you know how to handle, cook, and store food safely? Do you know how to take care of various foods and what to do with leftovers? If you understand the biology of the micro-organisms that can spoil food, you should now be able to answer "Yes!"

Working with Micro-organisms

To learn about the hard-to-see world of micro-organisms, you would begin by taking courses in microbiology at a university or college. For a career in medicine you could take courses on disease-causing micro-organisms. If you were interested in farming, you could learn about the micro-organisms in the soil and how they affect the way plants grow.

Many microbiologists work in the food industry. Every food industry—bakery, dairy, brewery, or meat-processing plant—needs well-trained people to check that dangerous micro-organisms are kept out of food. As well, they make sure that the useful micro-organisms grow properly. Some food-industry microbiologists have started using micro-organisms to help recover useful materials from food waste.

Doug Cunningham is a professor at the University of Guelph. His specialty is the use of micro-organisms in fermentation.

Doug co-ordinates a program in which students study for one term and then take jobs as microbiologists in food or other industries for the next. During their "work terms" the students earn

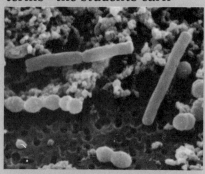

You might not see these fungi and bacteria while on a tour of a dairy, but you would certainly meet the microbiologists responsible for feeding and caring for them!

salaries and write reports on their work. When these students come back to class after their first months of work in the "real world," they are more interested in their studies. They now understand the sort of things that they need to learn in order to do the job.

Doug helps industry solve problems in microbiology. His work has taught him a lot about the many ways micro-organisms are useful in various industries. He passes on his knowledge to students interested in working as microbiologists.

Among the projects Doug and his students have tackled has been using micro-organisms to change the food waste left over after cheese-making into useful products. When you consider that a typical cheese plant may produce 23 000 L of food waste a day, finding ways to use it makes a lot of environmental sense!

TYPICAL CHEESE-MAKING

CHEESE-MAKING USING MICRO-ORGANISMS ON WHEY WASTE

milk

cheese

PASTEURIZED MILK

food for cattle

whey & yeast

cheese

whey & bacteria

methane fuel

whey (waste)

sewage plant

heat to pasteurize milk

Checkpoint

1. While preparing a meal one day, Winston performed the following actions:
   - washed his hands;
   - covered the extra food with plastic wrap;
   - put the extra food in the refrigerator;
   - heated the food on the stove.

   Explain how each of these actions reduces the risk of food spoilage.

2. (a) Name three ways of preserving foods that have been used for hundreds of years. Are these ways still used?
   (b) Do you think that modern methods of preserving food are necessary? Why or why not?

3. Many micro-organisms grow well where there is
   - moisture
   - warmth
   - oxygen
   - neutral (not acidic) conditions

   For each of these conditions, describe a method of packaging or storing that deprives micro-organisms of an environment that is favourable to them.

4. Describe one problem or disadvantage with each of the following methods of preserving food:
   (a) canning
   (b) use of chemical additives
   (c) freezing

5. A food distributor has one refrigerator truck and one regular truck. He has a shipment of locally grown strawberries to send to a town 150 km away and a shipment of fresh hamburger meat to send to a restaurant on the other side of town. Which truck should he use for which shipment? Explain your answer.

6. (a) Name two things connected with the food processing and packaging industry that are regulated by government (for example, labels).
   (b) State how each regulation you name helps the consumer.

7. You are going on a hiking trip for a week. You must carry milk with you, but you want to keep your backpack as light as possible.
   (a) List two alternatives to fresh milk.
   (b) State which type of milk and milk packaging you would prefer to take and why.

8. (a) What does this label mean?
   (b) Name three kinds of foods that would have labels like this on their packages.
   (c) What problem might you have if you used one of these foods after the time recommended on the label?

9. Suppose you want to buy some fruit and eat it right away. You cannot get the fruit washed at the store. Which fruit would you choose, an apple or an orange? Why?

Focus

- The main groups of micro-organisms are bacteria, fungi, viruses, protozoa, and algae.
- Most micro-organisms are single-celled, although some live in clumps and some are multicellular.
- Micro-organisms live nearly everywhere on Earth, including in the bodies of animals, plants, and in environments that would kill other living things.
- Micro-organisms can grow and reproduce rapidly and are found in large numbers.
- Micro-organisms perform essential functions in the environment and in the food chain, acting as decomposers, producers, and consumers.
- By decomposing human food, micro-organisms can cause spoilage and rotting; by fermenting certain foods, micro-organisms help produce and preserve them.
- There are many methods for protecting food from spoilage by micro-organisms, such as drying, freezing, and packaging.
- The various methods of preserving and storing food all have advantages and disadvantages, and food companies choose the method to suit the type of food and the people who will want to buy it.
- Standards and guidelines for food preparation help keep food free from contamination by micro-organisms.

- A system of government regulations and inspections helps ensure that the food offered to the public is fresh and safe.

Backtrack

1. Complete the word puzzle.

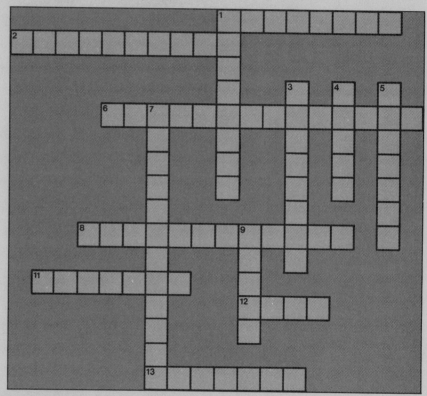

Across

1. The role of a micro-organism at the beginning of a food chain.
2. A term for an organism that helps to eliminate dead and waste material in the environment.
6. Living things that are studied with a microscope.

8. A process caused by micro-organisms that makes fruit juice bubble.
11. A method for keeping food from contact with air.
12. A substance used to grow a culture of micro-organisms.
13. Being free of micro-organisms.

Down

1. Micro-organisms that are single-celled, larger than bacteria, and could cause malaria or dysentery.

3. Micro-organisms that are single-celled, and may be spherical, rod-shaped, or spiral.

4. A micro-organism that is many times smaller than bacteria, and that must grow, feed, and reproduce inside a living cell.

5. A process for preserving food, especially meat and fish, that has been used for hundreds of years.

7. Things such as pesticides that accidentally get into food while it is being grown or processed and cannot be removed.

9. Yeast and mould are examples of this group of living things.

2. Explain the differences between
   (a) decomposing and spoiling
   (b) spoiling and fermenting

3. (a) List three different examples of methods used to preserve food.
   (b) Name one type of food commonly preserved by each example given.

5. Where might you find micro-organisms that grow well in the following conditions?
   (a) moist and salty
   (b) hot and dry
   (c) hot and wet
   (d) dark and without air
   (e) where there are piles of dead leaves

Synthesizer

4. Which term does not belong in each group? Write the term and explain why it does not belong.
   (a) consumer, producer, fermenter, decomposer
   (b) salting, smoking, drying, canning

6. "We'd be much better off without bacteria," said Trudy. "No more diseases or rotting food." "I'm glad there are bacteria," replied Eric. "We couldn't live without them." Who do you agree with, Trudy or Eric? Explain why.

7. Two types of micro-organisms were growing on a piece of bread. The bread was put in a freezer for one week. The bread was taken from the freezer and warmed up. One of the micro-organisms was dead, but the other continued to feed and reproduce. Why did one type of micro-organism die? How did the other one survive?

8. Fish can be bought fresh, smoked, frozen, or canned. Write down one reason why each method might be used in preference to the others.

9. Below are listed a number of stages during the process of getting vegetables from the farmer's fields to your dinner table. For each stage, name a method that can be used to help keep the vegetables free from contamination by micro-organisms. Explain how each method helps.
   (a) used when crops are growing in the field
   (b) used after crops are picked, before they are loaded on the truck
   (c) used when vegetables are being stored before processing
   (d) used during processing
   (e) used when processed vegetables are transported to the stores
   (f) used in the stores
   (g) used in the home

10. Melanie said, "It's the government's responsibility to make sure that the food we eat is safe." Do you agree or disagree? Identify the people and/or groups who you think are responsible for the safety of our food.

Changes on the Earth's Surface

The Earth is our home. Its mountains, hills, and plains seem to be a stable element in our rapidly changing world.

But these mountains and other landforms are not really as stable as they seem. The surfaces of the mountains in the first photograph are constantly being worn down and changed. The rocks in the second photograph became those curious shapes by a process of gradual change. Every year that mass of water rushing over the waterfall is eating away the rock underneath. And that blowing dust was once soil in which plants grew.

In this Unit you will learn about changes to the surface of planet Earth. You will explore mountains and valleys, rivers and lakes, glaciers and caves. You will see how even the hardest rocks are slowly broken into smaller pieces, and how these pieces are moved somewhere else by gravity, wind, water, and ice. All this change has brought about the landscapes we see today.

People have changed the landscape too. Just think of all the towns and cities that have been built on the surface of the Earth. Some changes made by people have resulted in problems. In this Unit you will think about such problems and about present and future solutions to them.

Changes on the Earth's Surfac

The Changing Face of the Earth

29 April 1903: FRANK, ALBERTA — A natural catastrophe has struck the citizens of Frank, in the scenic Crowsnest Pass. At 4:10 this morning an estimated 75 million tonnes of rock fell from Turtle Mountain, sweeping across the valley and 170 metres up the other side.

The roar of the slide was heard more than 150 kilometres away. The huge rockslide demolished part of the town. The coal mine underneath the mountain was almost totally buried.

Trees have been crushed, houses smashed, and the main railway line completely covered. Only the prompt action of a railway official saved the westbound train, the *Spokane Flyer*, from crashing into the debris in the darkness.

The landslide in 1903 at the townsite of Frank, Alberta.

Sudden Changes–Easily Seen

The Frank Slide changed the whole side of Turtle Mountain. A piece of the mountain about 700 m high, 1000 m wide, and 170 m thick, had fallen off the mountain. The slide also changed the valley: in less than two minutes from the start of the slide, millions of tonnes of rock had spread across the valley and up the other side.

After the slide, rocks continued to fall, and the survivors were afraid that there might be another slide. Eventually they abandoned the townsite, and Frank was rebuilt farther away. But the railway and road still run beneath the mountain. The remains of this disaster can be seen in the Crowsnest Pass, Alberta. It is a reminder of the dramatic changes that sometimes affect the Earth's surface.

More Sudden Changes

1. Study the photographs.
2. Explain what might be the cause of an avalanche.
3. (a) The sudden changes in photographs B, D, and E have caused certain problems for people. Describe the problem shown in each of these photographs.
   (b) Explain how the situations in photographs D and E might have been avoided.
   (c) Describe the steps people have taken to solve problems in photographs A and C.

Chicken wire fastened over a steep rock slope.

Structures called snowsheds have been built across the rail lines where avalanches occur most frequently.

Rocks are not the only things that fall off mountains. Where deep snow builds up, there is danger of snowfalls, or of **avalanches**—*(falls of snow, mud, and ice).*

Slower Change–Did You Notice?

A dramatic event like the Frank Slide produces noticeable effects. Most changes to the Earth's surface take place much more slowly. Some can be observed by anyone who visits the same place regularly over a period of time. Activity 6-2 shows how a student observed changes taking place around her school. Over the next year, watch for changes in your neighbourhood.

Activity 6-2

Changes During One Year

Joanne's junior high school opened just last year. She was asked to report on any changes she noticed around the school during the first year. Read Joanne's report, and then answer the questions that follow.

JOANNE'S REPORT

When we first came to school, the park around it was green grass everywhere. After a while the grass got worn where we walked most often. Soon there were paths across the park from the streets or alleys that led to the park. At first I wasn't sure if the paths were made by students. Now I do think the paths are made by students because they all lead to the nearest door of the school.

Before the snow came in November, the paths were all dirt. After the snow melted in spring, the paths got very muddy. We walked beside the mud and wore new paths beside the old ones.

Where the paths come over a hill, they have been cut deeply into the bank. I think this is because the rain runs down the path on the slope. I was going home once when it was raining, and I saw a muddy stream flowing down the steepest part. While I was there, a whole bunch of dirt slid down beside the path.

Finding Out

1. Joanne made two inferences in her report. What were they?

2. Which of Joanne's observations support the inferences?
3. Do you agree with her inferences? Explain why or why not.

Finding Out More

4. What might happen if the paths are used for many years without being paved?

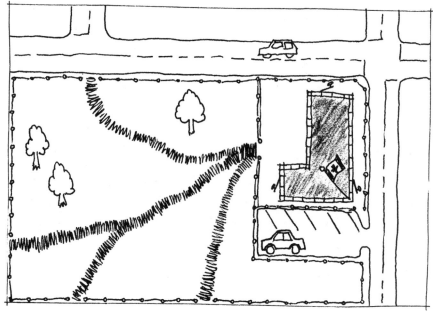

These are the paths that developed.

Slow Change—What Is the Evidence?

Most changes to the surface of the Earth take place much more slowly than the changes you have looked at so far. Slow changes are not easy to notice during the course of one person's life. How, then, can any person know that such change has taken place? Think about grass growing. You can't actually see it grow. But you don't have any doubt in your mind that it does grow when you have to cut it every week!

Now consider the slope in the photograph. You can't tell from looking at the photograph whether the broken rocks have fallen off the side of the mountain a few at a time, or in a sudden slide. If you watched the mountainside for a few days, you might see one or two pieces of rock fall and roll down the slope. A few pieces of rock falling off the mountain is not sufficient evidence to show that the change on this mountain is gradual and not sudden. But if you observed the mountain over many years, you might obtain such evidence.

Scientists who study the Earth are called **geologists**. Geologists have been observing the Earth over many years. Gradually the information they have gathered has begun to provide evidence for making inferences about what has happened—sometimes over many millions of years—on the Earth's surface. From this evidence geologists have suggested that slow changes over millions of years may have led to the rise and disappearance of landforms as large as mountains.

We do not have observations of the Earth taken over millions of years. But we do have other evidence that indicates change. You will be finding out about this evidence in the rest of this Unit.

5. (a) Observe change in your neighbourhood. Decide on a small area near your school or on your route from home to school. It might be part of a garden or grassy area on a slope, an area of earth beside the school wall, a river bank, or any other place where you think changes might occur within a period of a few weeks.
(b) Mentally mark out an area about one metre by one metre. Be prepared to watch for signs of change for a week, a month, or however long you can.
(c) In your notebook describe the appearance of the area during a dry period. Make a sketch if you can.
(d) Record what happens after a rainfall. Explain what might have caused any changes you observe.
(e) If possible, observe what happens in different seasons of the year, particularly at the end of winter after all the snow has melted.

Examples of Slow Change

This rock in Banff, Alberta, has been weathered over a long period of time. Read about weathering in Topic Two.

The Fraser River running through Hell's Gate Canyon, British Columbia. The canyon has been eroded by the fast-flowing river.

*These strange rock formations near East Coulee, Alberta are called **hoodoos**. Hoodoos have a "cap" of hard rock on top of a column of softer rock. The softer rock has been shaped by weathering and erosion. Read about erosion in Topic Three.*

Erosion in Dinosaur Park, Alberta.

Weathering–Wearing Down

In this Topic you will find out how the seemingly solid rock in mountains such as Turtle Mountain can be broken into small pieces. This process of breaking down rocks into smaller pieces is called **weathering**.

The most extensive weathering is brought about by physical forces, such as water, and changes in temperature. This kind of weathering is known as **mechanical weathering**. **Chemical weathering** is caused by the action of some chemicals. **Biological weathering** results from the activities of some organisms.

Mechanical Weathering

You found out in Unit Four that solids expand when they are heated. You discovered, for example, that railway tracks have gaps between sections to give the tracks room to expand on a hot summer's day. Rocks, too, expand on a hot summer's day, and contract again when they cool down at night or in cooler weather. Over a long period of time, the repeated expansion and contraction will cause the rocks to crack. Cracks are the first step in the process of weathering.

In the next step, water gets into the cracks in the rock. In cold areas—such as high up in the mountains, or in the cold of Canadian winters—the water in these cracks will freeze. Again over a long period of time, the freezing and thawing of the water in the cracks will break the rocks apart. This second step in the process of weathering is called **ice wedging**.

It may seem unlikely to you that an everyday process such as the freezing and thawing of water could help to break down huge mountains. The following Activity lets you test for yourself the power of water as it freezes.

Hot summers and cold winters will cause rocks like these to crack.

Weathering **303**

Ice Power

1. Find one or two small plastic bottles with screw tops. Fill them as full as you can with water and screw the tops on tightly. The containers represent cracks in rocks that have filled with water.
2. Place the containers in the freezer. You will need to leave them overnight or until the next lesson to find out what happens. Conditions in the freezer represent a cold night when water will freeze.

3. Take the containers out of the freezer and describe what you notice about them.

Finding Out

1. On the basis of the changes you observed in the containers, explain how water in cracks in rocks could break them apart.

2. The containers you used are much larger than many cracks in rocks. Explain why it would take a long time for water freezing and thawing in cracks to break rocks apart.
3. Why should you never store liquids in a glass bottle in a freezer?

The top layers of the mountain are gradually being worn down by weathering.

Imagine you work at a garden centre that sells large planters for growing flowers in the summer. You must explain to the people who buy your planters why they should empty them before the winter, or at least make sure the material in the planters is quite dry. In your notebook, write an explanation for the buyers of your planters.

Chemical Weathering

Mechanical weathering affects all rocks on the surface of the Earth. Some rocks are also broken down by chemical weathering. There are chemicals in the air such as hydrogen, oxygen, sulphur, nitrogen, and carbon. These chemicals may combine with moisture in the air or with rainwater to form weak acids. These acids gradually wear away at certain kinds of rocks. Test for yourself the effects of some chemicals on different rock types in the next Activity.

Fizzing Stones

Problem

Which rocks are affected by acids?

This Activity, or parts of it, may be done as a demonstration.

Materials

test tubes

test tube rack

safety glasses

safety apron

dilute hydrochloric acid

carbonic acid (contains carbon, from carbon dioxide, the gas that makes soda water and pop fizzy)

sand (formed by weathering from rock)

gravels and pebbles (may contain fragments of several rock types; you might test more than one piece)

granite chips (these or chips from other rock types may be available from a garden supply centre)

chalk or eggshell (these contain calcium carbonate, a chemical in both limestone and marble)

CAUTION: Acids are dangerous, even when they have been diluted. When handling acids observe safety rules. Take great care pouring any acid, and wear *safety glasses* and a *safety apron* while you do so. Wash any spills immediately with plenty of water and report them to your teacher.

Procedure

1. You will need a different test tube for each rock/acid combination you want to test. Set up your test tubes in a rack.
2. Give each test tube a number and record in your notebook which rock/acid combination you are testing in each test tube.
3. Put your rock samples in the numbered test tubes, and pour a little of the acid you are testing into the test tubes. Do not fill the test tubes more than halfway.

4. Observe and record what happens in each test tube.

Finding Out

From your observations, which types of rock do you think will be most affected by chemical weathering?

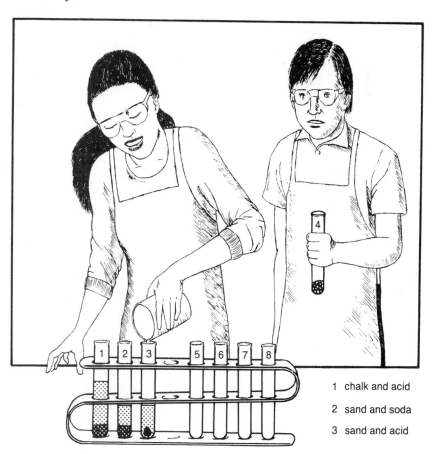

1 chalk and acid

2 sand and soda

3 sand and acid

Chemical Weathering and Landforms

Carbon dioxide in air combines with moisture in air or with rainwater to form carbonic acid. Limestone and marble are two kinds of rock that contain large amounts of calcium carbonate. In areas where there are limestone rocks on the Earth's surface, carbonic acid from the air has acted on the calcium carbonate in the limestone, and has formed large holes in the rocks.

Some of these holes form caves; the holes and caves may connect up and become underground tunnels. A river may disappear down a hole in the rock, flow in an underground tunnel for a while, and then reappear on the surface of the ground, perhaps many kilometres from the place where it disappeared underground.

How do we know this? Scientists have put coloured dye in the waters of such a disappearing river. If a river suddenly appears on the surface somewhere else, and its waters are the colour of the dye, then there is evidence that it is the same river.

A tunnel in limestone rock.

*Water drips constantly from various points in the roof of the cave. Calcium carbonate forms spikes that hang from the roof—called **stalactites** —and spikes that build up on the floor—called **stalagmites**.*

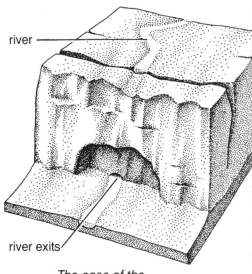

river

river exits

The case of the disappearing river.

[Comic illustration text]

y sorry, but our steps
take any more acid
ase use the trash
re entering.

Windsor
Castle
St. George's
Chapel

Drop a
Pop and
Save a
Step

Did You Know?

Castleguard Cave in Banff National Park is the longest cave so far discovered in Canada. An entrance to the cave was found in about 1920, but it was not explored until much later. Eighteen kilometres of tunnels had been explored by the mid-1980s. The far end, at a point 380 m higher than the entrance, has been found to be plugged with ice. Castleguard Cave was once estimated to be around 350 000 years old; new evidence now suggests parts of it may be over 10 million years old.

Acid Rain

In addition to naturally occurring chemicals, the activities of human beings add gases to the air. Such gases may come from the smokestacks of certain industries, or from automobiles. These gases join with the moisture in the air, or mix with rain, to form acids that return to the Earth as **acid rain**. Acid rain damages trees and slows plant growth. It enters lakes and rivers—sources of water for many living things. Chemical weathering from acid rain is seriously affecting many buildings.

The photographs show examples of chemical weathering. Many very old buildings have been gradually worn down over hundreds or even thousands of years by chemical weathering. Damage from acid rain is now causing much more rapid weathering of these structures.

Chichen Itza, Mexico.

Limestone and marble on buildings are affected by chemical weathering.

The Sphinx near Cairo, Egypt.

Weathering **307**

Biological Weathering

Have you ever seen weeds growing in cracks in broken pavement? Even a tiny plant can split pavement or rocks apart as it grows. Some tiny organisms called **lichens** grow directly on rocks. Their root-like structures let water into the tiniest cracks in the rocks. As the water freezes it widens the cracks. Lichens also make an acid that helps to break up the rock.

Animals that burrow in the ground may uncover rocks. These uncovered rocks may then be worn away by mechanical and chemical weathering.

Weathering resulting from the activities of living creatures is often called biological weathering. Biological weathering, as you can see, is usually accompanied by mechanical or chemical weathering.

This tree has pushed its roots into cracks in the rock. As the tree grows, its roots become thicker, and widen the cracks.

Lichens growing on rock that has been cracked by ice wedging.

Soils from Weathered Rocks

Without soils, many plants could not grow. And without plants, many other living things would not have food to eat. **Soils** are partly composed of loose, fine materials formed from rocks that have been weathered over many thousands of years. They also contain the remains of organisms that are decomposing.

Checkpoint

1. (a) Give one example of sudden change and one example of slow change on the surface of the Earth.
   (b) Give an explanation for each of these changes.
2. Imagine your teacher has asked you to do the following experiment. Fill a well slide with water and cover it with a regular glass slide. Hold the slides together with an elastic band and place them in a freezer. Examine the slides after the water has frozen.
   (a) Predict what your examination of the slides would show after the water has frozen.
   (b) Explain how this experiment might serve as a model for mechanical weathering.

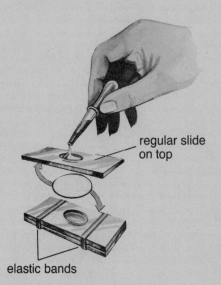

regular slide on top

elastic bands

3. Decide whether each of the following is an example of mechanical or chemical weathering, or both. Explain your answer.
   (a) Cracks in the sidewalk.
   (b) Discoloured metal coins found in a sunken ship in the ocean.
4. The coal mine underneath Turtle Mountain may have helped to cause the Frank Slide. The night before the slide it was very cold, with a heavy frost. Explain how weathering might also have helped to cause the slide.
5. The five diagrams show five steps as a pothole forms in the pavement. Read the description of each step, and then write a sentence to explain what is happening and why it is happening.
   (a) Rain is falling. Describe the condition of the pavement, and where the rain is going.

(b) It has turned very cold—well below freezing. What has happened to the ground under the pavement, and to the pavement? Explain why.

(c) A winter thaw has set in, and all the snow has melted. What effect has this had on the ground and on the pavement? Explain why.

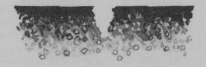

(d) Another bitterly cold winter day—it is freezing hard. What has happened to the pavement? Why has this happened?

(e) It's spring, everything has thawed, and the trees are in leaf. What has been the effect on the ground and the pavement? Why did this happen?

6. You have been asked to design a new city hall for an industrial city.
   (a) Give one example of a rock you would use as a building material in your design.
   (b) Give one example of a rock you would not use as a building material on the outside of the building.
   (c) What modern building materials might you use?

Erosion–Cutting Down and Carrying Away

The Frank Slide would not have occurred if the rock of Turtle Mountain had not been weathered. Yet, it was not weathering that caused the rock to come crashing down, but the pull of gravity. **Erosion**—the wearing away and movement of rock materials from place to place—may be caused by gravity, moving water, ice, or wind.

The Colorado River has cut its valley down from the top of the canyon.

The word **erosion** comes from a word meaning "to gnaw away" or "to eat."

Erosion by Moving Water

The Colorado River once flowed across a level plain of rock. As it flowed it gradually cut a valley in the rock, and some of the rock broke into pieces. The powerful river then carried these pieces away. Over many millions of years the river has carried away millions of tonnes of rock fragments, and cut a valley over one and a half kilometres deep. This valley is known as the Grand Canyon. It is one of the largest canyons in the world.

Modelling Moving Water

The Grand Canyon is a spectacular example of erosion by moving water. Streams and rivers are important causes of erosion, but they are hard to study directly. They may have water that is too muddy to see into. They may flood, dry up, or even change course. They may be thousands of kilometres long. And erosion caused by a river may take place too slowly for one person to observe even during a whole lifetime.

Scientists often use a model to help them study something, particularly when the thing to be studied is too small or too large, too costly, or too difficult to be studied directly. A useful way to study the behaviour of running water is to observe how it behaves in a simple model such as a stream table.

Streams run over different kinds of rocks. With a stream table you can observe the effect of water running over different types of rock materials. Streams run down mountainsides and across nearly flat ground. You can change the slope in a stream table and observe the effects that steepness of slope has on running water. Streams erode or move rock fragments of different sizes from place to place. In a stream table you can observe how running water moves rock fragments of different sizes.

In an actual stream, several of these conditions may occur at the same place. In a stream table you can observe each condition separately. For example, you can see the effect water running very slowly has on rocks of different sizes. Next you can make the water run faster and see what happens. Then you can increase the slope and see what happens. In the next Activity you will work with a stream table to test what effect water flowing slowly and quickly has on rock fragments of varying sizes.

Moving Rock Fragments

Problem

What effect does running water have on rock fragments of different sizes?

Materials

stream table
rock fragments of different sizes:
- sand (very small)
- gravel (small)
- pebbles (medium)

Table 1

SIZE OF ROCK PARTICLES	MY OBSERVATIONS
very small	
small	
medium	

Procedure

1. With your teacher's assistance, set up the stream table as in the illustration.
2. Mix the rock fragments and put them in the stream table to a depth of 3 to 4 cm.
3. Fill the bottom part of the table with water, to represent a lake.
4. Start by dripping water a drop at a time into the stream table. Then increase the flow of water to a gentle flow from the top of the table. The water represents a stream flowing over the rock fragments. Observe the effect of the moving water on the rock materials. Record your observations.
5. Predict what effect an increase in the speed of the water will have on
   - the sand
   - the gravel
   - the pebbles
6. Gradually increase the water flow to a medium speed. Observe the effect on the rock fragments and record your observations.

Finding Out

1. What happened to each size of rock fragment when water ran slowly?
2. What happened to each size of rock fragment when the water flowed more quickly?

Finding Out More

3. Predict what might happen to each size of rock fragment if you increased the steepness of the slope of the stream table.

Extension

4. Design an experiment to test the effects of moving water at different slopes on
   - sand
   - gravel
   - pebbles

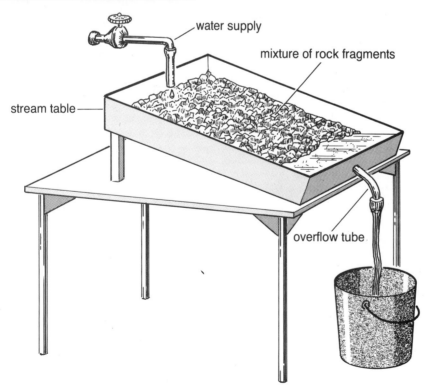

water supply
mixture of rock fragments
stream table
overflow tube

There's Gold in Them There Rivers!

Gold is found in many Canadian rivers, especially in the north and west. The **gold dust** (and sometimes even **gold nuggets**—lumps of gold) are mixed with sand, gravel, and pebbles in the rivers. Gold was first discovered in the West in the middle of the last century. Thousands of people joined in gold rushes to the remote areas where gold had been found. Travel was difficult, and the gold miners could take very little equipment with them. They had to have a way to separate the gold from the other rock materials in the rivers. They knew that gold is heavier than any other material likely to be found in a river. In the next Activity you will see how the gold miners used this idea to separate heavy materials, such as gold, from the gravel in the rivers.

Activity 6-6

Panning for Gold

Problem

How can heavier rock materials such as gold be separated from lighter ones?

Materials

stream table or large plastic garbage pail to catch the water and rock particles
gold pan (or metal pie plate)
gravel of mixed sizes, and sand
lead shot, or another heavy substitute for gold
water

Procedure

1. Mix the lead shot with the gravel in a pan.
2. Discard very large pieces of gravel that are obviously not "gold."
3. Half fill the pan with water, and swirl the mixture around. Observe what happens to the rock fragments in the pan.
4. Keep swirling the water and material in the pan, swirling out the lighter material over the edge of the pan into the stream table or garbage pail. Add more water from time to time as needed.
5. The "gold" will be visible as the other materials are removed. Keep swirling until there is mostly "gold" left.

Finding Out

1. Why were you able to separate the "gold" from the other rock fragments?
2. In Activity 6-5 you observed how rock fragments of different sizes were moved by running water. On the basis of your observations in Activity 6-6, explain how the different masses of various rock fragments might affect the speed at which they are moved by a stream.

Finding Out More

3. What forces, acting together, allowed you to find and remove the lighter substances after swirling them around in the water?

A Stream Story

The story begins with a rainstorm. Some of the rain that falls on open ground sinks into the ground. Rainwater that does not settle into the ground is called **runoff**. Runoff may cause erosion. Even a trickle of runoff flowing across loose soil will disturb the small pieces of soil. If the soil is on a slope, gravity and water together move the soil downhill. You may have observed changes caused by runoff in your neighbourhood.

Runoff

Soil erosion from runoff can be a serious problem in some areas. Steep banks and cliffs where few or no plants are growing are particularly liable to erosion from runoff. Even gentle slopes will lose soil in heavy rains. In some areas where people have cut down trees or ploughed under natural plants in order to grow crops, the soil has eroded away. People have been slow to understand the important part that plants can play in preventing soil erosion. Use the stream table to discover the effects of slope and growing plants on soil erosion.

Melting snow is another source of runoff. The water from the melting snow is eroding the loose surface materials in the road.

Activity 6-7

Soil Erosion

Problem

(a) What effect does the steepness of the slope have on the rate of soil erosion?
(b) Do growing plants prevent soil erosion?

Materials

stream table
soil, some with plants or grass growing in it

Procedure

1. Set up the stream table as in Activity 6-5. Put a layer of loose soil down one side of the stream table. Then put the soil with plants down the other side.
2. Set the stream table on a gentle slope. Let a small amount of water flow gently through each side of the stream table.
3. Record in your notebook what happens to each side of the stream table.
4. Increase the slope in the stream table, and repeat steps 2 and 3.

Finding Out

1. What effect did the steepness of slope have on the runoff on the bare soil?
2. What effect did the steepness of slope have on the runoff on the soil with the plants growing in it?
3. What differences were there in the effect of the runoff on each side of the stream table?
4. Explain the effect growing plants had on the movement of soil by runoff.

Streams

Rainwater runoff often flows into a stream. A small stream may start in a rocky, mountainous area where there is little or no soil and where the stream flows down a steep slope.

Rivers and Tributaries

As more and more streams flow down the mountain slopes, they join together to form rivers. Streams that flow into rivers are called **tributaries**. As more tributaries flow into the river, the amount of water in the river increases. Rivers flowing down steep slopes flow quickly.

Large Rivers

As the river flows into areas where the land is flatter, the river may spread out and become wider. The river now flows more slowly.

Rivers in Flood

When snow melts at the end of the winter, or when there are heavy rainstorms, runoff becomes much heavier. There is more water in the streams and rivers, which flow more quickly. Some rivers flood.

Activity 6-8

Stream Flow

1. Study the photographs above.
2. In your notebook draw a table with these headings: runoff, stream, river, large river, river in flood. For each stage:
   (a) record which describes the speed of the water: very slow, slow, medium fast, fast, very fast;
   (b) record which describes the amount of water: very small, small, medium, large, very large;
   (c) explain what effect the speed and amount of the water would have on the size and quantities of rock fragments that might be moved at each stage of the stream story.

Friction Plays a Role

As you have seen, running water moves rock fragments of various sizes, depending on the speed and the amount of water. Very large rock fragments are called **boulders**. Smaller pieces are called **pebbles**. Fine pieces are called **sand**. The finest pieces are muds called **clay** and **silt**. **Gravel** is the name given to rock fragments of various sizes between pebbles and sands.

In a fast-flowing stream, sand, silt, and clay are carried along in the water. Gravels and pebbles are bounced along the bottom. The sand, gravels, and pebbles wear down the river bed by friction as they are moved along it. The rock fragments are also themselves worn down as they rub against the river bed and against each other. Often the water wears the pebbles into rounded shapes with smooth surfaces.

Did You Know?

In a heavy rainstorm, rain droplets can fall on the ground at speeds up to 32 km/h. Each droplet makes a tiny crater in loose soil.

Water-rounded pebbles.

Rivers Cutting Down

At the beginning of this Topic there is a photograph of the enormous canyon cut by the Colorado River. Canyons or gorges have steep, straight sides, because the river has cut straight down into the rock below it. As time passes, the rocks at the top of these valleys become worn down by weathering. As a result, the valley becomes V-shaped. Rivers in canyons and in steep, V-shaped valleys are said to be younger rivers, because the sides of their valleys have not been worn down by weathering as much as the sides of valleys of older rivers.

This mountain stream is cutting down or eroding its valley as it flows along, making the valley deeper. Weathering is also wearing away the upper levels of the valley sides, making a V-shaped valley.

The Pounding Surf

The moving water in streams and rivers erodes the landscape over which it flows. The moving water in oceans, seas, and large lakes erodes the shorelines. Waves hit cliffs and shores like huge hammers. Loose pieces of rock picked up by the waves crash against the shore with great force. These rocks are broken down into smaller fragments as they are smashed together. As the tide goes in and out, waves wash over and erode different parts of the shoreline.

It takes a long time for the sea to erode some cliffs. But in other areas the rocks are easily worn away by the moving water, and erosion can be quite rapid. In England, one 50-km stretch of coastline is being eroded at the rate of about 2 m a year.

A large wave contains a great volume of water, and hits objects in its path with great force.

Water is constantly in motion along a shoreline.

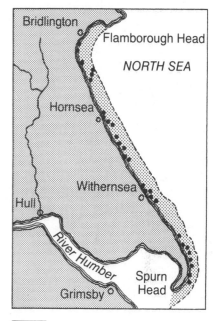

Bridlington
Flamborough Head
NORTH SEA
Hornsea
Withernsea
Hull
River Humber
Spurn Head
Grimsby

☐ land area

▨ land lost to sea

Five kilometres of the east coast of England have been eaten away in the last 2000 years. The dots mark places that were once villages but are now covered by the sea.

These isolated rocks were once part of the rocky coastline. The moving waters of the sea have worn away the rocks that joined these rocks to the coast.

Erosion 317

Blowing in the Wind

Rock particles are not only eroded by water. They may also be eroded by the wind. If the ground is moist and the wind light, plants can grow, protecting the soil from being blown away. But if the ground is dry and the wind blows strongly, plants cannot take root, and sand and soil are easily blown away. Extreme examples of wind erosion occur in the Sahara Desert in Africa and other sand deserts of the world.

Great Sand Hills, Saskatchewan. The wind has blown the loose sand on the surface into ripples.

The Dirty Thirties

When European farmers first came to the prairies in the early years of this century, the ground was largely covered with grasses of many kinds. The farmers ploughed huge areas to grow crops of grain. Ploughing left soil exposed to the wind. In the 1930s there was very little rain, and the soil dried out. Then wind blew the soil into terrible dust storms. The soil drifted like black snow, and often the grain seed blew away with the soil. When it did rain, the soil was washed away by the runoff. Because of these conditions, many farm families were unable to grow crops, and had to give up their farms and move away.

Despite their problems, the farmers joked about the effects of the wind. The cartoon on the next page is an illustration of one of the stories they used to tell. What facts are the basis of the story? Perhaps your family has some similar stories.

A farmer from the town of Skiff, on the Canadian prairies, remembers what it was like.

"The wind would howl at night, and soil would come through the ceiling like water. Every morning we would have to clean out the inside of the house. It just filled up with dirt again by the end of the day though. The fields were one big cloud of black dirt. All we could do was watch our fields blow away."

His son says that farmers today have learned from the problems of the past.

"We no longer plough huge fields, allowing wind to sweep across open areas of land. Rather, fields are planted in small strips and stubble is left on fields to help hold the topsoil down. Trees also provide protection to the fields."

As you have seen, the effect of moving water on rock fragments of different sizes varied with the amount of water and the speed it was flowing. Test what effect wind would have on soil pieces of different sizes. You can see from the photograph on page 318 that wind erodes sand, so you might select sand as one soil sample to test. For other examples, you might use samples that occur naturally, such as garden soil. Or you might want to use easily obtainable materials to represent soil samples that are fine and have a small mass, and soil samples that have a larger mass. For example, you could use uncooked rice or other grains or cereals.

Dust storm at Pearce airport, southern Alberta, April 1942.

Wind Erosion

Problem

What effect does wind have on particles of different sizes?

Materials

a hand-held hair dryer or a small fan

soil samples you have selected

a long, narrow, shallow container, such as the kind used in hanging wallpaper; or use a clear strip of floor. (Be prepared to sweep up at the end of the Activity!)

a plastic cup and some washers

Procedure

1. Put the samples together at one end of the container.
2. Turn on the hair dryer or fan to the slowest speed. Slowly bring the hair dryer close to where the samples are, and direct the "wind" at the samples so that it will blow them towards the other end of the container.
3. Record which soil samples move the farthest; which move the least. If possible, measure the distance each moved.
4. Gather the samples together again at one end of the container. Place an object such as a plastic cup with washers in it "downwind" (on the side away from where the "wind" comes) of the samples.
5. Repeat step 2 and record what happens.

Finding Out

1. (a) What effect did the size of the samples appear to have on the distance they moved?
   (b) What effect did the mass of the samples appear to have on the distance they moved?
2. Explain what happened to the plastic cup in step 5.

Finding Out More

3. Predict what would have happened to each sample if you had used the hair dryer at a higher speed.
4. What would have happened to each sample if you had dampened the samples before you used the hair dryer?

5. What would have happened if the samples had had plants growing in them?
6. From your observations of what happened to the plastic cup, explain why a farmer on the prairies might plant a row of trees near the farmhouse.

Extension

7. Design an experiment to test the effect of wind on damp soil samples, or on soils covered by growing plants. With your teacher's permission, carry out your experiment.

Did You Know?

People have started to try to grow plants at the edge of the desert. They hope that if they can get the plants to take root there, these areas of the desert will become available for growing food.

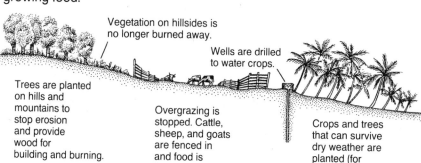

Vegetation on hillsides is no longer burned away.

Wells are drilled to water crops.

Trees are planted on hills and mountains to stop erosion and provide wood for building and burning.

Overgrazing is stopped. Cattle, sheep, and goats are fenced in and food is brought to them.

Crops and trees that can survive dry weather are planted (for example, date palms).

Sediments– Settling Down

You have seen that streams carry along in the water varying amounts of rock fragments. The amount depends on the size and mass of the fragments, the slope of the land the stream is flowing over, and the quantity of water in the stream. At some point along its course a stream slows down. It may happen that the stream's bed widens, and as the water spreads out, its speed slows. Or the stream may slow down because it is flowing across nearly flat ground.

Streams may also flow more slowly because they have lost some of their water. In warm, dry areas people may have used the water for farms, for industry, or for drinking water.

As a stream slows, it carries fewer rock fragments. The rock fragments the stream can no longer carry are set down, or **deposited**. These deposits are called **sediments**. The next Activity will give you some idea of what might happen to sediments of various sizes as they are deposited.

This is the Fraser River—the same river as in the photograph on page 302. Here the river has left the mountains and as it flows across flatter ground, it spreads out and slows down.

Settling Sediments

Problem

How do different kinds of rock fragments settle?

Materials

a tall transparent jar with lid
a mixture of fine gravel, sand, silt, and clay
water

Procedure

1. Write down your prediction of which sediment will take the shortest time to settle in the jar.
2. Fill the jar two-thirds full with water.
3. Pour the mixture of materials into the jar, shake the jar, and observe and record what happens to the various materials.

4. Leave the jar in a place where it will not be disturbed, and look at it again at a convenient time later, perhaps at the next science class.

Finding Out

1. How did the behaviour of the materials agree with your prediction?
2. How are the different sediments sorted in the jar? Try to explain why.

Finding Out More

3. A stream is carrying a mixed load of sediments out into a lake. What two forces are acting on the stream in the photograph as it flows into the lake? **Hint:** The stream is a moving object.
4. What will happen to the speed of flow as a result of these forces?
5. What will happen over time to the sediments you can see in the photograph?

A river carrying sediments as it flows out into a lake

Wandering Rivers

The water in streams and rivers is continually wearing away the ground over which it flows. When a river reaches its lowest level (that of the lake or sea it runs into), it cannot dig its channel any deeper. Instead it may start to move from side to side and make a curving bed. These curves are called **meanders**. A meandering river continually changes the landscape. You can use the stream table as a model to observe a meandering river.

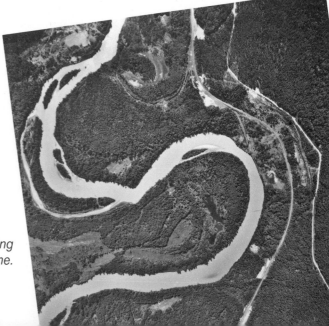

This photograph of a meandering river was taken from an airplane.

Wandering Water

Problem

What effects will meanders have on erosion and on the deposit of sediments?

Materials

stream table
2 plastic pails
40-cm length of rubber tubing
clamp
wet sand
eye dropper
food colouring
metre stick

Procedure

1. Before you begin making your model stream, sketch a meandering river in your notebook. Draw an arrow to show the direction the water is flowing, and make predictions to answer the following questions.

(a) Where will the greatest amount of erosion take place?
(b) Where might the river deposit some of this material?

2. Set up your stream table as shown. Make sure the sand is damp. Use your finger to make some shallow meanders in the sand.
3. Start the water flowing slowly into the stream table.
4. When the water is flowing at a slow, steady rate, add a drop of food colouring to the stream. Observe what happens both to the stream bed and to the food colouring as the water moves downstream.
5. Let the water run for 5 to 10 min, and continue to observe what happens.

Finding Out

Sketch the meandering river you have created in the stream table.
1. Show on your diagram:
(a) where in the stream bed the stream speeds up and where it slows down;

(b) where the greatest amount of erosion takes place;
(c) where the sediments are deposited;
(d) how the shape of the stream has changed over the period of time you made your observations.
2. How do your observations compare with the predictions you made at the beginning of the Activity?

Finding Out More

3. **Ox-bow lakes** are formed by meandering streams. Look at the ox-bow lake in the photograph and suggest how the stream may have formed it.

Extension

4. Try making an ox-bow lake in the stream table. Make the meander very broad, and run the water steadily. If the lake does not form, try varying the conditions in the stream table.

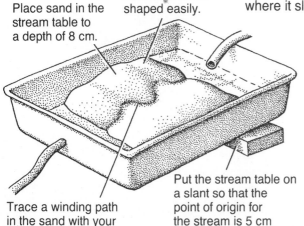

Place sand in the stream table to a depth of 8 cm.

Moisten the sand enough so that it can be shaped easily.

Trace a winding path in the sand with your finger to a depth of approximately 5 cm.

Put the stream table on a slant so that the point of origin for the stream is 5 cm higher than the end point.

River Floods and Sediments

When snow melts in the spring or when the rainfall has been particularly heavy, a river may flood, or overflow its banks onto the flat land beyond them. Rivers in flood carry large amounts of materials with them. As the water spreads out and its speed slows, it deposits sediments on this flat land. The sediments remain when the water recedes. These flat areas covered in sediments are called **floodplains**.

Deltas

Where a river flows into a very peaceful lake or sea, it slows almost to a stop. The river deposits most of its sediment at the place where it enters the lake or sea. Over time the sediments build up and form a triangular piece of land. This landform is called a **delta**, because it looks similar to the Greek letter delta Δ.

Sediments Moved by Seas and Lakes

Shorelines along seas and large lakes are changing all the time. The waves may be wearing away at the shore in one place, moving the rock fragments along, and setting them down somewhere else. Deep bays may be formed on one part of the shore where the water is slowly eating away at the land. Meanwhile, waves may be steadily building up new land at some other point along the shoreline.

In Lake Erie sediments have moved along the shore and have been deposited to form Point Pelee. Pelee Island, off Point Pelee, is the southernmost tip of Canada.

Probing

Use an atlas or a large-scale map of your province. Locate the largest lakes. Look for rivers flowing into them, and see whether there are any deltas in your province.

1. (a) Explain what runoff is.
   (b) What is the difference between erosion and weathering?
   (c) Give three causes of erosion.
2. (a) How does an increase in the slope of a stream's bed affect the speed of water flow?
   (b) How does it affect the quantity of rock particles the stream can move?
3. How does the amount of water in a river affect the quantity of rock particles the river can move?
4. (a) Sketch a meandering river.
   (b) Then sketch the course the river might take 50 years from now.
   (c) Explain why you have shown the river taking the course you have sketched.

5. When Marsha and Al Dirksen bought their farm, it had been abandoned for years. The ground was being eroded, and they were concerned about how they could make the farm productive again. The photograph shows the farm 15 years after they bought it. List the changes they made to stop erosion.
6. On the sketch of the meandering river you made in 4. (a), show the direction the river is flowing. Mark the section of the river where the water would be flowing
   (a) slowly;
   (b) fast.

7. Give two reasons why a stream table is useful as a model to study the effects of moving water.
8. The land in a floodplain makes particularly good farmland. Give reasons why this land is good farmland.
9. Should people be allowed to live on floodplains?
   (a) What are the disadvantages of living on a floodplain?
   (b) What are the advantages of living on a floodplain?
   (c) Should a river be prevented from flooding so that people can live on its floodplain?
   (d) How might floods be prevented? Try to describe more than one way.
   (e) Who should pay the costs of preventing the floods? The people who want to live there? All the people in the community? Have a class debate.
10. The conditions that caused the "Dirty Thirties" on the prairies could happen again. Tell why you agree or disagree with this statement.

Science and Technology in Society

Would You Construct a Building Here?

Imagine that you have just been asked to design and build a magnificent new convention centre for Edmonton, Alberta. It's to be located on Grierson Hill overlooking the North Saskatchewan River valley.

As an architect, you are excited and challenged by the job, but you are also concerned about the proposed site. You know that the hill is steep and unstable. There was a massive landslide in the early 1900s, and there have been many other smaller ones since then.

You also discover that 50 years earlier there were coal mining operations on the hill. The mining added to the instability of the hillside.

So the problem you are faced with is this: How do you construct a large building on the side of a hill without weakening the hill further and possibly triggering more landslides?

This was, in fact, the challenge faced by the engineers, architects, and geologists who designed the Edmonton Convention Centre. The first step, they agreed, was to study the hillside in more detail.

First, core samples of the soil and underlying rock were taken. These samples revealed many zones of

Erosion of the hill at the Grierson Hill Road site before construction began.

Shafts were dug into the side of the hill to mine coal.

"weakness" on Grierson Hill, some right under the proposed site of the building! A "tiltmeter" was installed to measure any movements occurring on the hill. Other instruments were used to determine the groundwater conditions.

The studies suggested that a building could be constructed on the site, but only if the hill were "stabilized." Before you read on, see if you can think of some practical approaches to stabilizing a hillside.

The engineers thought they could slow down the erosion of the hill by removing a portion of the upper slope. But they knew that simply removing the soil might weaken the hillside even further. Before any excavation could begin, an extremely strong retaining wall had to be built. This wall was made of metre-wide steel cylinders that were driven deep into the hillside, well below the level of excavation. These cylinders were held in place by steel anchors cemented into the surrounding earth and rock.

But once the building was constructed, how would anyone be able to tell if the wall and anchors were continuing to do their job? Instruments called "load cells" were implanted into some of the anchors to monitor the surrounding soil. The steel anchors could then be adjusted if there were any changes in the soil that might cause a shift in the load on the wall.

The Convention Centre was built right into the side of the hill. This was done to ensure extra stability. The Centre extends well beneath Jasper Avenue, which runs across the top of the hill.

In the end, the work to stabilize the hill cost 7 million dollars, about a third of the total cost of building the Centre. It was an enormous undertaking, but now thousands of visitors to Edmonton can enjoy a breathtaking view of the river valley below, thanks to the efforts of the people who designed and built the Edmonton Convention Centre.

The Edmonton Convention Centre under construction.

The beginning of construction on the Grierson Hill site.

The completed centre as it looks from the river.

Water Underground

The stream story in Topic Three started with rainwater running off the soil. Not all the rainwater that falls on the soil becomes runoff. Some of it sinks into the ground. This water is called **groundwater**.

The water that sinks into the ground enters the surface layers of soil. Does the water stay mixed with the soil near the surface, or does it continue to sink farther down? You have found that soil is partly composed of finely weathered rock. Just as there are many kinds of rock, so there are many different types of soils. Does water behave differently in different soil types? If water sinks through soil, where does it end up? The next Activities let you think about these questions.

WHACK!

NORMAN EYOLFSON '89 ©

Dripping Down

Problem

How quickly does water flow through different soil samples?

Materials

jar for catching water
empty frozen juice can with 4 or 5 holes in the base (or an empty plastic pop bottle with the base removed and a filter of cheesecloth fastened to the base with elastic bands)
graduated cylinder
stopwatch or clock with second hand
dry gravel
dry sand
dry potting soil
dry peat moss
dry clay
water

Procedure

1. Half fill the can with gravel.
2. Measure 50 mL of water in the graduated cylinder.
3. Pour the water quickly into the can with the gravel. Record the time from the moment the water starts to drip until it has stopped dripping.
4. Empty, rinse, and dry the can, and repeat the procedure for each of sand, potting soil, peat moss, and clay.

Table 2

MATERIAL	TIME TAKEN FOR 50 mL OF WATER TO DRIP THROUGH
gravel	
sand	
potting soil	
peat moss	
clay	

Finding Out

1. Through which soil sample did the water flow most quickly? most slowly?
2. Suggest reasons for any differences in the time taken for each material.

Extension

3. Predict how long it would take for water to drip through a mixture of equal amounts of clay and gravel; a mixture of equal amounts of soil and sand.
4. Make these mixtures and test your predictions.
5. Some house plants, such as cacti, come from environments where there is very little water. What kind of soil mixture would you use to grow these plants? Explain why.
6. Some house plants come from very moist environments. What soil mixture would you use to grow these plants? Explain why.
7. From your observations, explain why clay alone is a poor soil for growing any kind of plants.

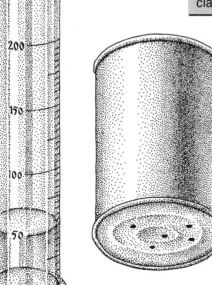

The Water Table

In Activity 6-12 you poured water onto different soil substances. There were spaces between the particles in the soil samples. The water was able to flow into these spaces. Before you poured the water in, these spaces were filled with air.

When water falls on the ground, it works its way down through the spaces in the soil. Eventually it reaches rock that does not have spaces large enough for the water to pass through. As rain continues to fall, the level of the water above this rock rises. This level is called the **water table**.

Between rainfalls all plants—the crops in the fields, the trees in the woods, and the grass in the backyard—rely on the water in the soil. They take in this water through their roots.

Water on the Surface

Where the water table reaches the surface of the ground, there are lakes and sloughs. Where the water table reaches the surface in a valley, there are streams and rivers. In wet weather the level of the water table is high, and there is plenty of surface water. Lakes, sloughs, streams, and rivers are all examples of **surface water**. If there is a drought, the level of the water table falls. Sloughs and shallow streams are likely to dry up. Plants cannot get the water they need. That is why many plants die during a period of drought.

Rain is not the only source of water for the water table. The melt from winter snows raises the level of groundwater. Too little snow in winter can lead to droughts that affect the crops in the summer.

Activity 6-13 provides a simple model for observing how rainwater might form groundwater and various kinds of surface water. By pouring water onto sand, you can watch the effect rain has when it falls on dry ground. By making holes or hollows of different depths in the sand, you can see how water fills sloughs, lakes, and wells.

Water in the Well

Materials

shallow container that will hold
  water, minimum 30 × 20 × 5 cm
  (large baking tray)
open-ended plastic tubing
sand
water supply
beaker or paper cup

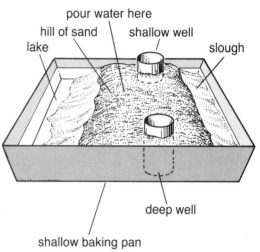

Procedure

1. Half fill the container with dry sand, leaving a slightly uneven surface.
2. Clear one corner to the bottom of the container ("lake").
3. Make a shallow hollow on the other side of the container ("slough").
4. Stand two tubes into the sand and make sure they remain upright. Try to keep the inside of the tubes as clear of sand as possible. Push one nearly to the bottom of the container, and the other only about 2 cm into the sand ("deep and shallow wells").

5. Pour a little water very slowly from the beaker onto the sand in the container. Observe what happens to the water.
6. Continue to pour until water appears in one of the holes in the container. Note where the water first appears.
7. Continue to pour water gently into the container. Note when water appears in the holes you have made and where it appears on the surface.

Finding Out

1. When water was first poured into the sand, where did it go?
2. Which hole first showed water?
3. Where did water appear next?
4. Where did water first appear on the surface of the sand?

Finding Out More

Based on your observations of this model of the behaviour of groundwater, explain the behaviour of the groundwater in Johann's and Cheryl's camp.

5. When the rain fell in the spring, what happened to it?
6. How did water get into the wells, streams, and slough?
7. Why was one of the wells dry when Cheryl visited the camp?
8. What happened to the slough in the summer?

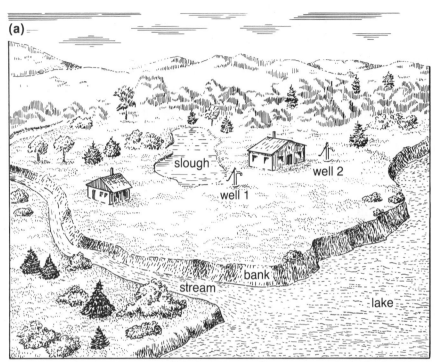

(a)

Johann spent a holiday at a camp in the country one spring. It rained a lot that spring, and there was water in both wells, and in the slough, the lake, and the stream.

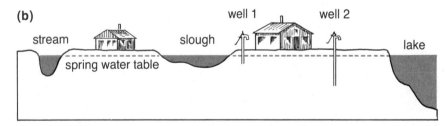

(b)

This diagram shows the ground under the camp in the spring, when Johann visited it. You can see the level of the water table and where the water table reaches the surface.

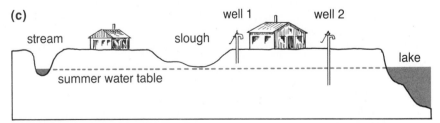

(c)

This diagram shows the level of the summer water table when Cheryl visited the camp.

Glaciers

Topic Two described how very small amounts of ice can weather large rocks and break them into pieces. Just imagine what large amounts of ice might do. Great ice sheets have changed the landscape over large areas. They have eroded the surface and deposited large quantities of sediments. Where did these huge ice sheets come from?

In land areas at the extreme north and south of the world near the poles, and in high mountain regions, it is so cold that the snow never melts, even in summer. Each winter new snow falls on top of the snow that remains from the past summer. Each year, then, fresh snow builds up on top of the older layers of snow.

Snow is heavy. Remember what it feels like when you have to help shovel it? As the layers of snow get very deep, the bottom layers are squeezed together. Sometimes when you make a snowball, if you squeeze really hard the snow seems to turn to ice. That is what happens to the bottom layers of snow. The weight of the snow as it piles higher and higher packs the lower layers into ice.

Think of the way pancake batter or muffin mix spreads out in a pan as you pour. In a similar way, as the ice builds up higher and higher, its weight causes it gradually to spread out. These large, spreading sheets of ice are called **glaciers**.

High up in these mountains the snow never melts, even in summer.

The huge Athabasca Glacier moves down the valley from the Columbia Icefield in western Alberta. Visitors to the glacier can walk on the ice or even take a bus across it.

Continental Glaciers

The two largest sheets of ice in the world are the glaciers on Greenland and Antarctica. These two ice sheets are called **continental glaciers**, because each covers a large area of land. In Greenland the ice is 3 km thick; it covers all of the land except a narrow strip near the coast. In Antarctica, where temperatures reach some of the lowest recorded on Earth, the ice is nearly 5 km thick in places; it covers the whole continent and even extends out into the sea.

Rivers of Ice

High in the mountain ranges snow piles up year after year. It packs down into ice and the ice spreads out just as it does in continental glaciers. When the ice reaches the top of a steeply sloping mountain valley, gravity causes the ice to move down the valley, just as gravity causes a river to flow down a slope. Glaciers that move down from the mountains in this way are called **valley glaciers**. They have also been called rivers of ice because they flow.

This diagram shows the area of the ice sheet on Greenland and how it spreads in all directions.

How a glacier is formed in a mountain region.

Snow builds up year after year. The bottom layers turn to ice.

The snow and ice spread out.

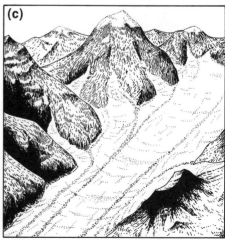

When the snow and ice reach a slope, gravity pulls the glacier downhill.

Moving Forward and Melting Back

As a valley glacier moves down the mountain slopes, it reaches lower levels where it is warmer. Here snow melts in summer. So does the ice at the tip of the glacier. The tip of the glacier where the ice is melting is called the **snout**.

A river ends when it flows into a lake or the sea. The glacier ends at the snout. The position of the snout is not at the same place every year. In some years it may move farther down the valley. When this happens, the glacier is **advancing**.

The glacier is always moving down the valley, so you might think that the snout would always be advancing. But remember that the ice is also melting at the snout. So the combined effect of moving down the valley and melting does not always mean that the glacier is advancing. The diagram will help to explain this.

Sometimes the snout stays at the same position for a number of years. This happens when the ice is melting at the same rate as the glacier is moving forward. The glacier is **stationary**. And in some years the snout may even move farther up the slope of the valley. When this happens, the glacier is **retreating**.

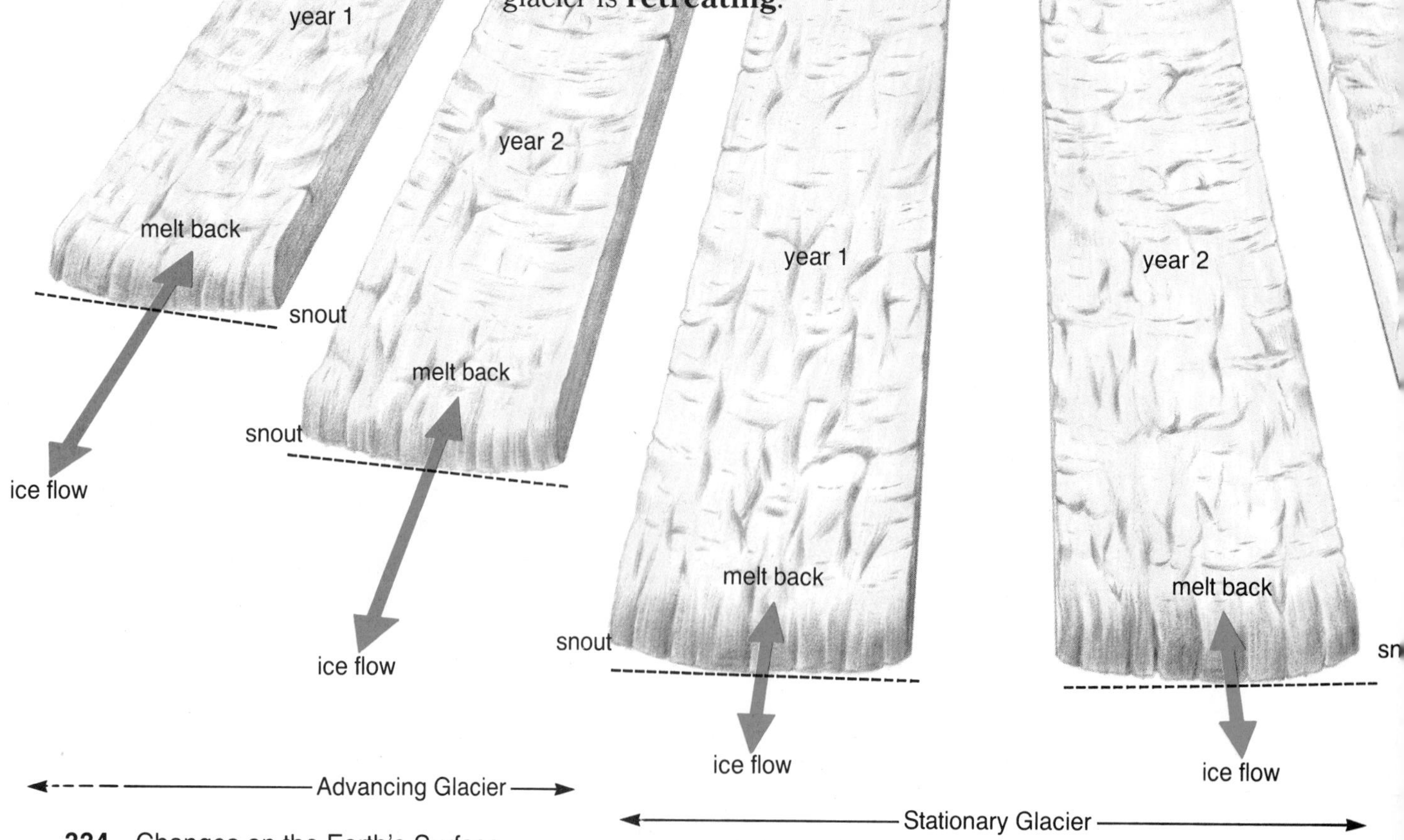

year 1

melt back

snout

ice flow

year 2

melt back

snout

ice flow

← - - - ——— Advancing Glacier ——→

year 1

melt back

snout

ice flow

year 2

melt back

sn

ice flow

←——— Stationary Glacier ———→

The Puzzle of the Boulders

About a hundred and fifty years ago geologists in northern Europe noticed examples of erosion and deposition that could not easily be explained by the action of streams and rivers. For example, they found boulders of a rock type totally different from all the other rocks in the area. These boulders were too large to have been carried any distance by the normal flow of a river. In some places they found deposits of pebbles and gravels that were also of different rock types than those in the surrounding area. They noticed that the rock types from which the boulders, pebbles, and gravels came could often be found many kilometres farther north.

*The Big Rock near Okotok, Alberta, is an **erratic**—a boulder, left behind by a glacier, of a different rock type from the other rock in the area.*

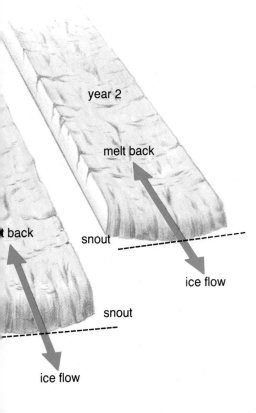

year 2

melt back

back

snout

ice flow

snout

ice flow

— Retreating Glacier - - - - ➤

Two European geologists, Johann van Charpentier and Louis Agassiz, were studying the Earth's surface in Switzerland, where there are many valley glaciers. They noticed certain kinds of deposits and signs of erosion that the glaciers had made as they advanced and retreated in the valleys. They compared these features with the signs of erosion and deposition that other geologists had found in areas where there are no valley glaciers. They wondered whether *all* these features might not be the effects of moving sheets of ice, or glaciation, across the land many years in the past.

You can use ice cubes to represent glacier ice in order to observe some of the effects ice might have had on the landscape.

Ice Studies

Problem

(a) What effects might rock fragments in glacier ice have on the surface of the Earth?

(b) What might happen to rock fragments in glacier ice as the ice melts?

Materials

sand
gravel
several ice cubes
several ice cubes made from water mixed with gravel
several ice cubes made from water mixed with coarse sand
stream table
piece of cardboard

Procedure

1. Set up the stream table and put a layer of sand 3 to 4 cm deep in it. Bury several clear ice cubes in the sand. Also bury ice cubes with sand and gravel in them. Wait for the ice cubes to begin to melt.
2. While you are waiting, pull a clear ice cube across the piece of cardboard. Record any effects on
   (a) the cardboard;
   (b) the surface of the ice.
3. Pull an ice cube with gravel frozen in it across the piece of cardboard. Record the effects on both the cardboard and the ice surface.
4. Repeat the procedure for the ice cube with sand frozen in it.

5. Now go back to the stream table.
   (a) Observe and record what happens to the sand around the ice cubes as the ice melts.
   (b) Observe and record what happens to the gravel and sand that was frozen into the ice as the ice melts.

Finding Out

1. What effect do rock pieces in the ice have on the surface over which the ice moves?
2. When ice melts, what happens to the rock fragments that were frozen in the ice?
3. As the **meltwater** flows from melting ice, what channels does it make in the sand?

Ice Across the Land

Charpentier and Agassiz noted that many of the features they had observed in areas of valley glaciers were similar to physical features on the Earth's surface a long way from the valley glaciers. They began to look at more and more signs of the effects of moving ice in areas far from valley glaciers. They found enough evidence to suggest that sheets of ice had once covered large areas of Europe over a period of many years. Geologists have since found similar evidence in North America that sheets of ice have covered large areas of Canada and the northern United States. The periods when ice sheets covered areas of Europe and North America are known as **Ice Ages**.

Valley Glaciers

There are many glaciers in mountain ranges in western Canada. A close look at the effects these glaciers have had on the landscape could help in interpreting the evidence for or against past glaciation.

Like rivers, glaciers carry loose rocks and stones along with them as they move. Some of these have fallen onto the glacier from the surrounding mountainsides. Others are dragged from the valley sides and floor and moved along the valley bed by the bottom layer of ice in the glacier. As this bottom layer moves along with its load of rock fragments, it scratches the rock and the ground beneath it. These scratches in rocks are called **striations**.

As the ice pushes along the walls of the valley, a ridge forms along the sides of the glacier, caused by the build-up of rocks there. You can see these ridges, called **lateral** or **side moraines**, in the photograph. Rock pieces in a glacier also scratch the sides of the valleys. Where the ice melts at the snout, the rocks and particles are deposited on the ground. They form a ridge, called an **end moraine**.

End moraines form ridges along the snout. When a glacier stops in one place for a time, the ridge becomes higher and higher. When a glacier retreats, there will be ridges from end moraines at each place the snout pauses.

Side moraines deposited by the melting glacier form ridges running in the same direction as the glacier is moving. You can see a high ridge of lateral moraine on the right of the photograph.

Ice Ages—the Evidence

What you see in the photographs here and on the previous pages is only part of the evidence scientists have found to support the theory that there were Ice Ages in the past. Here is a list of some other evidence that supports the theory that there were once Ice Ages. As you read this list, bear in mind that the glaciers probably advanced in the general direction from north to south.

- Deposits of rock fragments from end moraines and meltwater form a ragged line across the North American continent. This line marks where the ice sheets ended—the farthest point south that they moved.
- **Drift**—the sediments deposited by glaciers—covers large areas of northern United States and Canada north of the line of moraine deposits.
- There are many more lakes north of this line than there are to the south; geologists infer that the glaciers eroded many of the basins (depressions) these lakes occupy.
- Eskers, drumlins, kettle lakes, and striated rocks are found only north of this line.

This diagram shows some of the features that are left behind by a glacier as it retreats across the land. **Outwash** *deposits are left by the meltwater of the glacier.* **Braided streams** *are formed as meltwater from the glacier runs in shallow channels over loose sediments.*

Eskers: *long, winding ridges of glacial deposits. Water melting below the ice in the glacier runs in tunnels under the glacier. These tunnels become partly filled with sands and gravel, which are deposited when the glacier melts.*

Drumlins: *long, smooth, tapering hills formed from drift. Drumlins are usually found in clumps, all pointing in the direction the glacier moved. Drumlins were probably formed when a glacier ran over material deposited in a moraine by an earlier glacier.*

drumlins

ice sheet

esker

erratic

outwash

braided stream

kettle lakes

drift

outwash plain

end moraine

Striations: *long, parallel scratches on rocks resulting from rock fragments carried along in the ice.*

Kettle lakes: *depressions in the outwash plain formed when buried blocks of ice melted.*

Glaciated valleys: *as a glacier flowed along a valley, the rock fragments it carried gouged and scratched the sides of the valley.*

The Ice Ages–Past and to Come?

Scientists have evidence that at least four ice sheets advanced and then retreated across North America and Europe within the last million years. During these Ice Ages it was much colder than it is today. The retreat of the last ice sheet was about 11 000 years ago. We are now in a warm period. Scientists do not know if, or when, another Ice Age might come.

Some scientists think that the increasing number of people in the world and the effects of the technology we use in modern life could bring a period even warmer than the present one. If such a warming trend does occur, the glaciers could retreat still farther. Scientists are continuing to study the evidence for and against future Ice Ages.

On the basis of the evidence, geologists suggest that the glaciers in North America may once have covered all the shaded area on this map.

Did You Know?

The tops of the Cypress Hills in Alberta and Saskatchewan were not completely covered with ice during the last Ice Age. Some of the plant and animal species living today in the Cypress Hills are different from those in the surrounding lowlands. These differences might have arisen if these species lived on the Cypress Hills for a long time when ice covered the surrounding land.

Available freshwater is becoming scarce, and people have been looking for ways to increase the supply. Australians have considered towing icebergs from Antarctica to their countries. An iceberg melting in the warmer waters around these countries would provide a large supply of fresh water.

Icebergs

Some of the glaciers in the Arctic flow down to the sea. They continue to flow forward, even though there is no land to support the ice. As it is too cold for the ice to melt at a snout, great chunks break off the unsupported glacier and form **icebergs**. This process is called **calving**.

Some of these icebergs are huge—many times larger than the largest ocean-going ship. Because of their size, icebergs can be very dangerous to ships.

Icebergs calving into the seas from Portage Glacier, near Anchorage, Alaska.

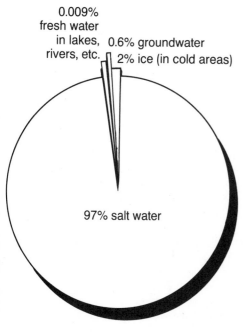

0.009% fresh water in lakes, rivers, etc.

0.6% groundwater

2% ice (in cold areas)

97% salt water

Three Uses for an Icefield

Every time you drive on a road or travel on a train you can thank the glaciers. They left eskers and deposits of gravel that provide the materials used to make roads and railbeds.

The only freshwater available for people to use is in groundwater, and in surface water such as rivers and lakes. The only other source of freshwater in the world is the ice in the glaciers and icefields. If a warming trend did occur, some of this ice would melt and become available for people to use.

Icefields in the Rockies are a major source of water for people living on the prairies. Some of this ice melts every summer and ensures a continuing supply of water in the rivers, whether it rains in the summer or not.

Working Outdoors

Rob Babiarz is a forest resource technologist. "Most people think that all forest technologists do is fight fires. Fighting fires is only a small part of my job."

What then does a forest technologist do? Rob Babiarz says, "We plan the harvest of our forests in a way that will have the least impact on the environment and at the same time give the best economic returns."

When a company wants to log a particular area, it submits a logging plan. The plan describes the location and size of the trees the company wants to cut, and the logging roads it will need to build. Rob and his co-workers study the proposal to make certain that there will be little or no damage to the environment. They tell the company where, how, and what the company may log. For example, instead of cutting down all the trees, the company might only be allowed to cut small groups of mature trees. Or the company might be required to leave a large strip of forest around the edges of a stream or a lake to prevent erosion and provide shelter for wildlife.

Rob's job also involves checking on the company while it is logging. "On some jobs, we go into a forest area on a Monday morning and stay there until Friday afternoon." For anyone who likes working outdoors, a career in forestry has lots of appeal.

What made Rob decide on his career? Rob worked for two summers in the forestry service and became convinced that this was the career for him. He enrolled in the Forest Resource Technology program, where he studied soil science, weather, physics, and botany.

Interested students can join Junior Forest Ranger or Warden organizations during the summer. They take part in planting tree seedlings, observing logging operations, and investigating the relationships among plants and animals in the forest. Joining an organization like Junior Forest Rangers is a good way to find out whether you would like a career in forestry.

1. Solve the crossword puzzle by using words from Topics Five and Six.

2. (a) Give two examples of surface water.
   (b) Explain why the water table may be at different levels at different times.
   (c) Explain why plants may die during a long drought.

3. (a) What is the difference between a continental and a valley glacier?
   (b) Write two sentences to explain the advance and retreat of glaciers.

4. Brad visited a very interesting area last summer. He thought the area might once have been covered by glaciers. What clues could have led Brad to make this inference?

5. (a) List three ways in which glaciers erode the landscape.
   (b) List three types of glacial deposits and describe the kinds of materials they contain.

6. In many rural areas, people get their water from wells. As more people build holiday homes in the countryside, more and more wells are dug.
   (a) What problem is likely to result?
   (b) Describe two possible solutions to the problem.
   (c) Evaluate your two solutions, and decide which you would choose. Explain why you chose that solution.

7. In many parts of the world, water is in short supply. Water conservation is especially important in these areas. One of the main goals of people living in these parts of the world is to increase the amount of groundwater. Why is an increase in groundwater important?

8. Explain why rocks high up in mountain valleys might be striated.

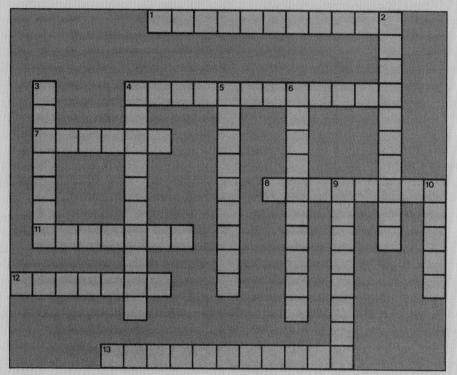

Across Clues
1. rain that sinks into the soil
4. lakes, rivers, sloughs, and streams are this
7. long, winding ridges deposited by glaciers
8. long, smooth, tapering hills formed by a retreating glacier
11. a large spreading sheet of ice
12. the process by which icebergs are formed from glaciers
13. depressions formed when buried blocks of ice melted

Down Clues
2. a glacier moving farther up the slope of a valley is _____
3. a great chunk broken off a glacier
4. scratches in rock caused by moving glacier
5. a glacier moving down a valley is _____
6. the level of water in the ground
9. side and end _____
10. the melting tip of the glacier

Unit Review

Focus

- Some changes on the Earth's surface are sudden, such as rock slides and avalanches. But most changes on the Earth's surface take place very slowly.
- Weathering is a process of wearing down the Earth's surface by mechanical, chemical, or biological means.
- Rivers and streams erode the land across which they flow. They carry boulders, pebbles, sand, gravels, and silt along with them.
- Moving water in seas and lakes erodes rock at shorelines. Wind erodes dry, open ground by blowing soil or sand away.
- Sediments are rock particles that have been moved from their original place and deposited somewhere else. They may be clay, silt, sand, gravels, pebbles, or boulders.
- As streams flow more slowly, they deposit sediment. Sometimes they form a delta.
- Groundwater is rain and meltwater that sinks into the ground. The water table is the level below which the spaces in the ground are filled with water.
- Glaciers are formed in areas where snow gathers from year to year and is compressed into ice. Glaciers flow; the ends of a glacier may advance, and retreat. Glaciers cause changes in the landscape by erosion and deposition.

Backtrack

1. (a) Give a definition of these words:
   avalanche
   delta
   groundwater
   erosion
   sediment
   weathering
   meander
   deposit
   glacier
   (b) Write a sentence about rivers using at least two of these words.
2. Explain the processes of mechanical, chemical, and biological weathering. Give an example of each.
3. How do rivers and streams erode the landscape?
4. Explain how a delta is formed.
5. Why may the water table be at different levels in wet and dry periods?
6. Give four examples of how glaciers alter the landscape.
7. Explain how "panning for gold" is an example of a good way to separate sediments.

Synthesizer

8. Rachel and José were hiking in the Rocky Mountains. They came across an old stone building that was no longer in use. Plants were growing in parts of the stone floor. The walls were full of cracks. Explain why the walls and floor were in this condition. Explain what might happen to the building in the future and why.

9. People make changes in the environment.
   (a) Make a list of six changes to the landscape in your neighbourhood that have been caused by the actions of people.
   (b) Beside each change, state whether the change might result in any harmful effects. Suggest what might be done to avoid such effects.
   (c) List changes that have had or that might have helpful effects.
10. Imagine that you have moved into a new house or apartment. Around it is some sloping ground without any plants. Plan a hillside garden for this area. Be sure your plan includes your reasons for decisions you have made.
11. Pictures of the surface of Mars show boulders and blowing dust. Yet the first human footprint on the Moon's surface is still there, many years after it was made. Why do you think the surface of Mars and the Earth have more in common than the surface of the Earth and the Moon?
12. You are going to build a holiday cottage in a beautiful mountain canyon. The sales agent has shown you five possible sites for the cottage. Each site is the same price. The sites are shown in the diagram.
    (a) Draw a table in your notebook similar to the one shown.

(b) Read the sales agent's description of each site carefully.

(c) Study the diagram, and then fill in the table.

(d) Evaluate the information in your table. Which site would be your first choice? Explain why you chose that site.

(e) You find that the site has already been sold. Which is your second choice? Explain why.

Site 1

The site is on a high bank overlooking the river. You will have an excellent view. There is a steep rocky cliff directly above the site—just great for you to build directly against the cliff. You can clear away the loose rocks at the foot of the cliff, and build on the rock beneath it.

Site 2

The site is on a gentle slope. It will be easy to get down to the river for boating or swimming. The building site is on rock.

Site 3

This site is on a steep hill, giving a good view. You could clear away the loose rock on the slope down to the river below the site, and make a path.

Site 4

A nice site close to the river level. In spring you'll have a tributary river right beside your cottage. There is a gentle slope down to the river for easy access.

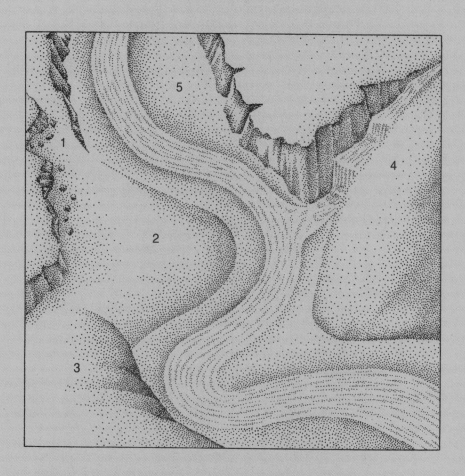

Site 5

At the river level. Direct access to the river—only a few steps to go boating. A lovely site on river sediment of sand, gravel, and pebbles.

SITE	ADVAN-TAGES	DISADVAN-TAGES
1		
2		
3		
4		
5		

Skillbuilder One

Units of Measurement

Measuring

Have you ever estimated a distance by pacing it off? Sometimes pacing off gives a reasonably accurate estimate of a distance. For example, if you were with a group of students studying wildlife on a field trip, you might need to pace off distances in the area you were studying. But you would have the same person do all the pacing. Why would you need to do that?

If other people wanted to study the wildlife in the same place, they would need to know the exact distances that were paced off. The students would therefore need to compare the length of the pace they used with a standard unit of length.

As you probably know, the standard unit of length that we use is the metre. A metre stick, a tape measure, or any other measuring device with a **scale** marked in equal divisions makes it possible for you to make exact measurements. A measuring instrument with a scale, even one as simple as a metre stick, can come in handy. A scale helps you find out how many units there are as you measure, without having to count them one by one.

Some SI Units of Measurement

The system of measurement that we use in Canada today is the Système internationale d'unités (SI). This system was developed in France in 1791. It was revised and modernized in 1960 and is now used in over 100 countries throughout the world. We often refer to it as the metric system; the name comes from the base unit of length—the metre.

Table 1 shows the units of measurement that are used in this book.

Table 1 *SI Quantities, Units, and Symbols*

QUANTITY	UNIT	SYMBOL
length	metre*	m
	centimetre	cm
	millimetre	mm
	kilometre	km
mass	kilogram*	kg
	gram	g
	tonne	t
volume	litre	L
	millilitre	mL
time	second*	s
temperature	degree Celsius	°C
force	newton*	N
energy	joule*	J
	kilojoule	kJ

* SI base unit.

SI is easy to use and understand because all the units of measurement are derived from the base units by multiplying or dividing by ten. The other units show their relation to the base units by the prefixes in their names.

Table 2 *Some SI Derived Units*

QUANTITY MEASURED	PREFIX: MULTIPLE:	milli $\frac{1}{1\,000}$	centi $\frac{1}{100}$	deci $\frac{1}{10}$	UNIT 1	deca 10	hecto 100	kilo 1 000
Length		millimetre mm	centimetre cm	decimetre dm	metre m			kilometre km
Area			square centimetre cm²		square metre m²			
Volume (solid)			cubic centimetre cm³		cubic metre m³			
Volume (liquid)		millilitre mL			litre L			kilolitre kL
Mass		milligram mg			gram g			kilogram kg
Energy					joule J			kilojoule kJ

Science Skills

Why is an infant's head so large?
Why does a bridge need supports?
Why do things move?
Why does boiling water turn to steam?
Why does food spoil?
Why do avalanches occur?

Science enables you to explore the world around you by asking questions like these, and finding ways to answer them. In **Science Directions 7** you will find answers to these and many other questions. In doing so, you will use these science skills:

- observing
- classifying
- estimating
- measuring
- predicting

- making inferences
- stating a hypothesis
- designing and doing experiments
- recording and organizing information
- analysing and interpreting data

Observing

In your science class, you'll use your senses of sight, hearing, touch, and smell to make observations. It is important to record your observations as you make them.

Classifying

You **classify** objects or organisms by grouping together those that are alike in some way. You can group things in many ways: by size, colour, shape, or other characteristics.

RELAX MOM... IT'S JUST A SIMPLE MATTER OF CLASSIFICATION

NOW!

Estimating

When you swing a bat at a ball, you estimate the speed of the ball and judge the moment you should swing your bat. You learn to make estimates of many kinds—time, distance, speed are just a few—by practice.

Measuring

Words like big and small, or hot and cold, do not give exact information about height, length, mass, or temperature. Scientists try to avoid such vague words. Instead, they use precise words and careful measurements. Read *Skillbuilder Three* to learn more about measurement.

THEN THIS HUGE DOG BARKED AT ME!

WOOF!

ARF!

Predicting

Predicting is forecasting what might happen, based on previous observations and experience. You can therefore give reasons for a prediction you make. You can test your predictions by doing experiments.

Making Inferences

An **inference** is a possible explanation for an observation or series of observations. Sometimes you can make more than one inference from the same observations. You can test the accuracy of some inferences by doing experiments.

Stating a Hypothesis

A hypothesis is an idea or an explanation, based on previous observations and experience, of *why* something might *always* happen.

Experiments

When you do an experiment you plan and carry out a series of activities that enable you to:

- find an answer to a question
- solve a problem
- observe how something behaves

Experiments can test predictions, inferences, and hypotheses.

Experiments involve some or all of the science skills you've read about in this *Skillbuilder*. Which of the skills might Jane have used to invent her "wake-up" machine?

Variables in Experiments

In many experiments you make observations over a period of time. When you plan your experiment, you decide for how long you will make your observations. For example, you may make observations every 30 s, every 5 min, or once a day. In other words, you regulate the time measurements in your experiment. The measurement or condition that is regulated by the person doing the experiment is called the **manipulated variable**. (Something that varies, or changes, is a variable.)

The **responding variable** is what you measure to find the data you obtain from your experiment. For example, you may want to know how fast a plant grows from the time the first shoot appears above the surface of the soil. You decide that you will measure the plant's growth once every three days for three weeks (the time is the manipulated variable). You measure and record the plant's growth in centimetres: the plant's growth is the responding variable.

Controlled Experiments

Suppose you want to find out what conditions affect the growth of a plant. You have noticed that a plant growing in a pot on a window sill grows well. The plant gets plenty of sunlight. You make a hypothesis that *plants need sunlight in order to grow well.*

You decide to test your hypothesis to find out if sunlight is important for plant growth. In order to test the need for sunlight, you would have to grow at least two plants:

(1) give one plant water, good soil, and sunlight.
(2) give the other plant water, good soil, but no sunlight.

The second plant is your **control** in the experiment. It should have exactly the same conditions as the first plant, except for the sunlight. Without that plant as a control you could not learn whether a lack of sunlight alone affects the growth of plants. Experiments in which the experimenter keeps all factors (variables) the same except for one, are called **controlled experiments**.

Recording and Organizing Information

As you work through **Science Directions 7**, you will be recording a great deal of information (data). You will record observations you make during activities; predictions you want to test; and probably some ideas of your own.

There are a number of ways you can record and organize information.

(1) water, good soil, and sunlight

(2) water, good soil, but no sunlight

Title: Investigation 5B.4
The Bouncing Tennis Balls

Problem: How does the bounce height of
Brand X tennis balls compare
with other brands?

Prediction: I think that Brand X
tennis balls will bounce
higher than all other
brands when dropped from
a height of 1 m.

Materials: metre stick, 3 different
brands of tennis balls.

1 m

Procedure:
1.

My Force Meter

①. Works best when I pull things.

②. Needs stronger resistance to test
larger forces.

Table of Temperature Changes				
Time (min)	Estimated Temperature (°C)	Recorded Temperature (°C)	Changes Predicted	Changes Observed
0				
1				
2				

*After you have read **Skillbuilder Four**, return to this page.
List as many ways as you can to improve the graph shown
here.*

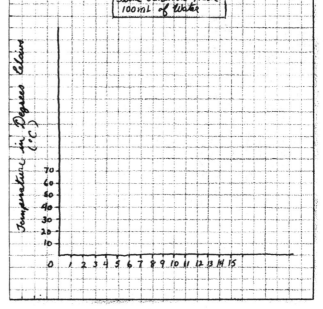

stalk

veins

*Notes, tables, graphs, and sketches are all ways of
recording and organizing data.*

Analysing and Interpreting Data

You usually need to analyse the data you have obtained from an experiment to find out whether there is a pattern or trend in the data. The pattern or trend may provide an explanation for your results. Often graphs will help you to see patterns or trends (see *Skillbuilder Four* for more about graphs).

Sometimes you need to analyse data even further before you can make any inferences from them. The Finding Out questions at the end of the Activities in this book will help you to interpret your data.

Doing a Science Project

The information in this *Skillbuilder* and the organizers that follow will help you to plan and carry out a science project of your own. Use the diagrams shown here to remind yourself of the steps you might follow as you draw up a plan for your project.

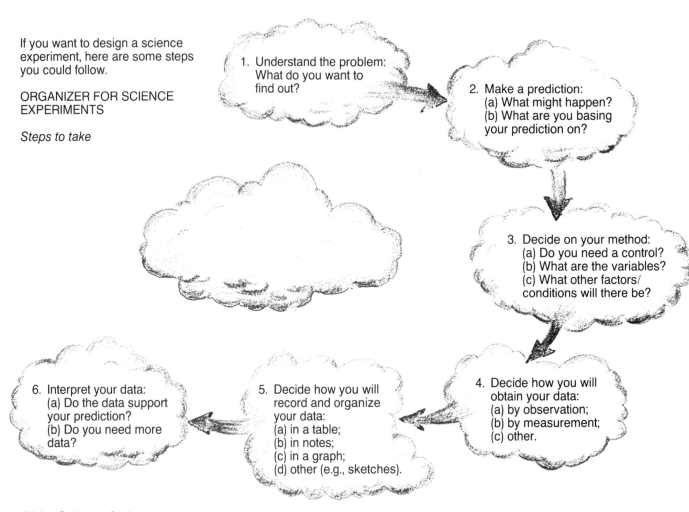

If you want to design a science experiment, here are some steps you could follow.

ORGANIZER FOR SCIENCE EXPERIMENTS

Steps to take

1. Understand the problem: What do you want to find out?

2. Make a prediction:
   (a) What might happen?
   (b) What are you basing your prediction on?

3. Decide on your method:
   (a) Do you need a control?
   (b) What are the variables?
   (c) What other factors/conditions will there be?

4. Decide how you will obtain your data:
   (a) by observation;
   (b) by measurement;
   (c) other.

5. Decide how you will record and organize your data:
   (a) in a table;
   (b) in notes;
   (c) in a graph;
   (d) other (e.g., sketches).

6. Interpret your data:
   (a) Do the data support your prediction?
   (b) Do you need more data?

If you want to solve a practical problem by inventing or designing and building a model, you might want to follow these steps.

ORGANIZER FOR SOLVING PRACTICAL PROBLEMS

Steps to take

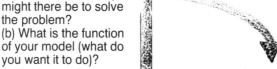

1. Understand the problem:
   (a) What different ways might there be to solve the problem?
   (b) What is the function of your model (what do you want it to do)?

2. Develop a plan:
   (a) Make a checklist of things you need:
   • materials;
   • shapes of the parts;
   • how to fasten any joints.
   (b) Draw a sketch or diagram of your model.

4. Evaluate your model:
   (a) Does it work?
   (b) Did it solve the problem?
   (c) Can you now see a better way to solve the problem or to build your model?

3. Build the model and test it.

If you want to investigate an issue that affects people, think about following these steps.

ORGANIZER FOR ISSUES

Steps to take

1. Understand the issue:
   (a) Write it in your own words.
   (b) Decide what alternative points of view there are.

2. Find information about each point of view:
   (a) Research to find information.
   (b) Organize the information.
   (c) Decide what information supports each point of view.

6. Evaluate the alternative solutions to the issue.

3. List the alternative solutions to the issue:
   (a) Do any of the solutions affect people?
   (b) Do any of the solutions have an effect on the environment?
   (c) Do any of the solutions cost money?
   (d) What might be the results of each solution?

7. Decide whether you need any additional information.

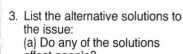

Measuring

Measuring Length

The SI units of length are the kilometre, the metre, the centimetre, and the millimetre. Use whichever unit is most appropriate for the length you are measuring.

Skillbuilding Practice 3-1

Which Unit?

1. Which unit would you use to measure how high you jumped in a track and field event?
2. Which unit would you use to measure the width of this book?
3. Which unit would you use to state the distance from your city or town to the mountains?
4. Describe a situation when you would want to measure in millimetres.

Estimating

Estimating measurements can often be very useful. Are you good at estimating length? Look at these illustrations, and notice the length each unit represents.

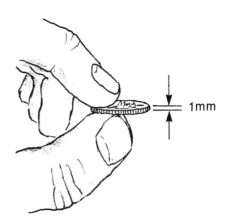

A millimetre is about the thickness of a dime.

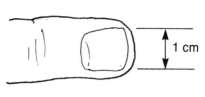

A centimetre is about the width of a fingernail.

A metre is about half the height of a door.

How Long?

1. Look at the following list of classroom objects. Using a table similar to Table 1, record your estimates of the length, width, or height of each object. Use a suitable unit of measurement for each.

Table 1

OBJECT	ESTIMATE	MEASUREMENT

(a) the thickness of an eraser
(b) the length of a pencil
(c) the width of a desk
(d) the height of a chair

2. (a) Now measure and record the length of each of these objects, using the same unit of measurement as you used in your estimate. Record your measurements in the table.
(b) How close were your estimates to your measurements?

3. (a) Record on the chalkboard several students' estimates of the width and length of your classroom.
(b) Measure the width and length of your classroom and compare the measurements with your classmates' estimates.

4. You have probably been using a ruler or a metre stick to make your measurements. Try to think of an instrument that is used to measure distance in kilometres. Describe this instrument.

Measuring Mass

Mass is the amount of matter in an object. The mass of a building is made up of the materials used to build it—concrete and glass, for example—as well as the people inside the building. Your body's mass (which is measured in kilograms) is made up of your bones, teeth, hair, skin, and all the other things your body is made of.

Have you ever tried to estimate the mass of some objects? If you have, test yourself. If not, practise now.

How Much Mass?

1. Estimate the mass of:
   (a) an apple
   (b) an empty beaker
   (c) your notebook
   (d) your wristwatch
   Record your estimates in a table similar to Table 2.

2. Compare your estimates with those of other classmates.
   (a) Did the estimates differ? By how much?
   (b) Which unit of measurement did most of you use for each estimate?

Table 2

OBJECT	ESTIMATE OF MASS	MEASUREMENT OF MASS

Instruments for Measuring Mass

The illustrations show some common instruments for measuring mass.

A classroom balance scale, used for measuring mass accurately.

Supermarkets use electronic scales to measure mass.

Skillbuilding Practice 3-4

Measuring Mass

1. Examine the balance scale you use in your classroom. Study the illustration to learn the parts of the balance.
2. Measure the objects whose mass you estimated:
   (a) the apple
   (b) the empty beaker
   (c) your notebook
   (d) your wristwatch

   Record your measurements in Table 2. How accurate were your estimates of mass?

3. (a) What is your mass?
   (b) What do you use to measure your mass?
4. Suppose you want to measure the mass of water in a beaker.
   (a) First measure the mass of an empty beaker and record it.
   (b) Fill the beaker half full of water.
   (c) Then measure the mass of the beaker containing the water and record it.
   (d) Work out and record the mass of the water.

Measuring Volume

The volume of an object or substance is the amount of space it occupies. As you work through **Science Directions 7** you will be measuring the volume of substances in litres and millilitres.

What things are measured by mass, and what things are measured by volume? Think about the things you would find at home in your kitchen cupboards and refrigerator. Would milk, jam, sugar, flour, salt, pickles, and fruit juice be sold by mass or volume?

Skillbuilding Practice 3-5

Mass or Volume?

1. Look in your kitchen at home and fill in as many items of food as you can in a table similar to Table 3.
2. What kinds of foods are sold by mass (kg)? What kinds of foods are sold by volume (L)? Explain the main difference between the foods sold by mass and those sold by volume.

Table 3

FOOD MEASURED IN KILOGRAMS (kg)	FOOD MEASURED IN LITRES (L)
Flour	

Measuring Liquids

The volume of a liquid can be measured accurately by using a container with a scale marked on the side. This kind of container is called a graduated cylinder ("graduated" means "divided into equal parts").

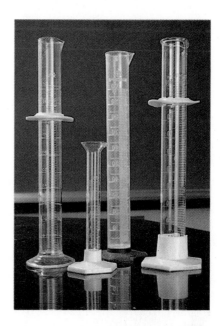

Several sizes of graduated cylinders are available. Plastic rings are placed around glass cylinders to help prevent them from breaking.

The surface of a liquid in a glass or plastic container is not perfectly flat. The liquid often rises slightly up the sides of the container, and the level of the liquid dips slightly in the centre. That dip in the surface of the liquid is called the **meniscus**. You must look at the level of the liquid in the *bottom* of the meniscus to measure the volume of liquid accurately. The best way to do this is to look at the surface of the liquid at eye level.

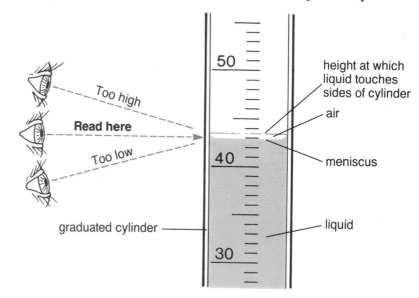

How to read the level of the liquid in a graduated cylinder.

Volume

1. Pour any amount of water into a jar or other container. Try to estimate its volume. Record your estimate.
2. Pour the water into a graduated cylinder and record the volume.
3. Repeat steps 1 and 2 several times. Can you improve your ability to estimate volume?

Reviewing Measurement

1. Estimate and measure in suitable units:
   (a) the length of a TV screen
   (b) the mass of a banana
   (c) the volume of water in a partly filled jar
2. Give an example of an object that is about
   (a) 1 cm long
   (b) 50 cm long
   (c) 1 km long
3. Give an example of something that has a mass of about
   (a) 1 g
   (b) 50 g
   (c) 1 kg

4. What instrument would you use to measure the distance you cycled in one hour?
5. (a) A friend wonders if a soft drink really does contain exactly 280 mL of pop. Describe the steps your friend should take to find out.
   (b) You are not sure whether an unopened box of cereal contains the mass that is printed on the outside of the box. How would you find out if it does?

Skillbuilder Four | Graphing

When you are doing science activities, it is important to observe carefully, measure accurately, and record your data in your notebook, either in notes or in a table. Often your notes and tables contain clear evidence on which you can base inferences. But sometimes the record of your observations is so complicated that it is not easy to analyse the data. In these cases, you may need to reorganize your data before you can interpret them.

One useful way to reorganize a set of numbers in a table is to make a graph of the data. A **graph** shows numerical data in the form of a diagram.

There are three kinds of graphs that we commonly use: bar graphs, line graphs, and divided circle graphs. Each kind of graph has special uses. **Bar graphs** are particularly helpful when you want to compare (1) characteristics of different objects, (2) quantities of different substances, or (3) amounts of something and changes in that amount. Bar graphs, in other words, help you to make *comparisons*.

Line graphs (1) show changes in measurement, or (2) make it possible for you to decide whether there is a relationship between two sets of numbers. Line graphs may help you to see whether "if this happens, then that happens."

Divided circle graphs enable you to see what share or proportion of a whole various items represent. Divided circle graphs are also called **pie graphs**. Pie graphs are not only round like a pie; they show you how large a share of the pie is represented by each object or substance illustrated on the graph.

Bar Graphs

It can be difficult to make comparisons between numbers in a list. Look at the following table.

By scanning the table carefully, you could pick out the highest and the lowest of these ten mountains. But it would be hard to picture how much higher the highest mountain is than the others. If you made a bar graph, you could see quickly, for example, which mountain is the second-highest, how much higher it is than the third-highest mountain, and so on. Here's how to draw a bar graph.

Draw graphs on graph paper. To graph the height of the mountains, ten bars are spaced evenly along a horizontal line towards the bottom of the graph. Because each bar represents a different mountain, there must be a space between each bar and the next. The vertical line up the left-hand side of the graph must have a scale. This scale represents the heights of the mountains. Leave space on the left of your graph for this scale. Leave space below your graph to label each bar with the name of the mountain it represents.

Choosing a scale for a graph takes some skill. You decide what scale to use by looking at the **range** of the numbers, that is, the lowest and highest numbers in the data. In Table 1 the lowest mountain is Ruwenzori at 5109 m, and the highest mountain is Mount Everest at 8848 m. Scales are usually rounded off to whole numbers. So, your scale must range from 5100 m to 8900 m, or better, from 5000 m to 9000 m. A scale should always extend beyond the numbers in the data.

When you have chosen the range of your scale, you must decide what **intervals**, or numbered divisions to mark on the scale. For example, you could not mark a division for every metre on a scale that ranges from 5000 to 9000 m.

To graph the data in Table 1, should you use a scale marked in intervals of 1000 m? Look at Table 1. Would you be able to work out where to mark the height of Ojos del Salado on such a scale?

Table 1 *The Ten Highest Mountains in the World*

MOUNTAIN	HEIGHT IN METRES (m)
Kilimanjaro, Africa	5895
Ojos del Salado, South America	7084
Ruwenzori, Africa	5109
Mount Everest, Asia	8848
Mount McKinley, North America	6194
Mount El'brus, Europe	5642
Kanchenjunga, Asia	8598
Huascaran, South America	6768
Mount Logan, North America	5951
K2, Asia	8611

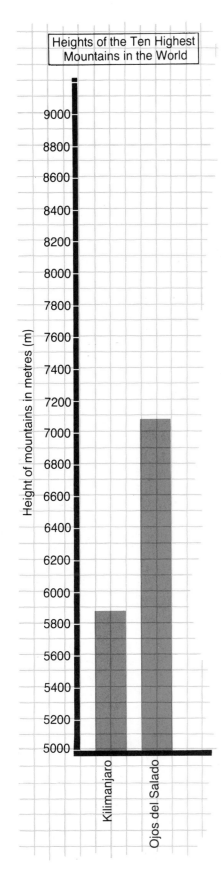

Heights of the Ten Highest Mountains in the World

Height of mountains in metres (m)

9000
8800
8600
8400
8200
8000
7800
7600
7400
7200
7000
6800
6600
6400
6200
6000
5800
5600
5400
5200
5000

Kilimanjaro

Ojos del Salado

Should you use a scale marked in intervals of 100 m? Would it be easy to mark the height of each mountain on such a scale? Calculate how many 100 m intervals there are between 5000 and 9000. Count the squares on your graph paper. Will a scale with this many intervals fit on your graph paper?

Decide which interval will work best on your scale on your graph paper.

Once you've chosen the scale and marked it on the vertical line on the left of your graph, you are ready to plot the points—the height of each mountain—on the graph. Study the illustration to see how to do this, and complete your graph. Note that when the numbers in your data fall between the divisions marked on your scale, you have to estimate where to plot the point.

Check yourself.

- Have you written on your graph the scale and the units of measurement used in the scale?
- Have you labelled each bar to show which mountain it represents?
- Have you given your graph a title to tell what your graph is about?

Now use your graph to answer the questions in the Skillbuilding Practice.

Skillbuilding Practice 4-1

Reading from a Bar Graph

1. Which is the highest mountain? the third-highest?
2. Which is the lowest mountain? the third-lowest? What is the difference in height between them?
3. Are the mountain heights evenly spaced across the scale, or do there seem to be groups of mountains with similar heights? Explain, mentioning the names and heights of the mountains.
4. Could you have answered these questions as easily from the table as from the graph?

Drawing a Bar Graph

Draw a bar graph of the data in Table 2.

1. Decide on a scale, and mark it on your graph.
2. Plot the points on the graph, and draw the bars.
3. Label the bars to show what each represents.
4. Give your graph a title. Draw a block at the top of your graph paper. In the block, print the title, your name, and the date.
5. Which is the longest whale?
6. By how much is it longer than the next-longest whale?
7. What is the range between the longest and shortest whales?

Table 2 *Average Length of Different Kinds of Adult Whales*

KIND OF WHALE	LENGTH IN METRES (m)
Blue	30
Bowhead	18
Finback	25
Grey	14
Humpback	15
Minke	10
Right	17
Sperm	18

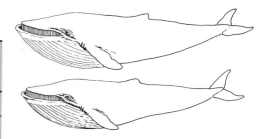

More Bar Graphs

1. Explain what the bars on this graph represent.
2. Why do you think the data have been grouped in these bars?
3. How many students had a mark of
   (a) 60-64?
   (b) 85-89?
   (c) 55-59?
   (d) 90-94?
4. Do you think the scale chosen for this graph has appropriate intervals? Explain why or why not.
5. Look at Table 3. Which group of animals has the most kinds, or species?
6. Could you show all this information on one bar graph?

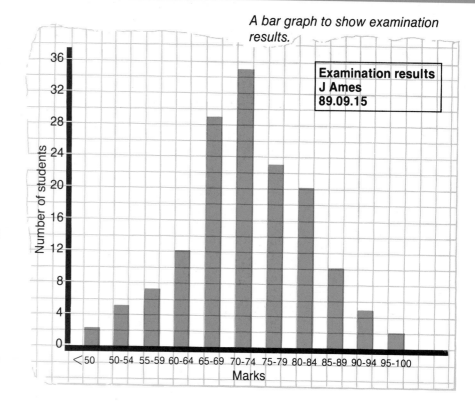

A bar graph to show examination results.

Examination results
J Ames
89.09.15

Line Graphs

Line graphs are made up of two number lines. The horizontal number line is called the **x-axis**. The vertical line, called the **y-axis**, starts at the left end-point of the x-axis. The point at which the x-axis and the y-axis meet is called the **origin**. The illustration shows how to draw the origin and the two axes (axes is the plural of axis) on graph paper. Note that you do not draw either axis close to the edge of the paper. You should leave some space to print a scale and a label along each axis.

The manipulated variable is displayed on the x-axis, and the responding variable is displayed on the y-axis. In the line graph shown below, the depth of the bath water is the manipulated variable (the one you change) on the x-axis. The y-axis shows the responding variable—the one that changes when you change the depth of the bath water, (Read *Skillbuilder Two* if you need to review variables.)

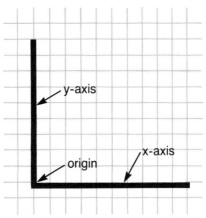

The x-axis and the y-axis meet at the origin.

Skillbuilding Practice 4-4

Reading a Line Graph

1. Suppose you like your bath water as deep as possible. In most bathtubs the maximum depth is about 25 cm. How much would that bath cost you?
2. How much could you save by running a bath that was only 20 cm deep?
3. How much could you save by running a bath that was only 10 cm deep?
4. How much would you save in one year by having one bath each day that was only 10 cm deep instead of 25 cm deep?

Table 3 *Numbers of Kinds (Species) of Animals*

ANIMAL GROUP	NUMBER OF KINDS
fishes	20 000
amphibians (frogs, salamanders)	3 000
reptiles (snakes, lizards)	6 000
birds	8 600
mammals	4 300
insects	800 000

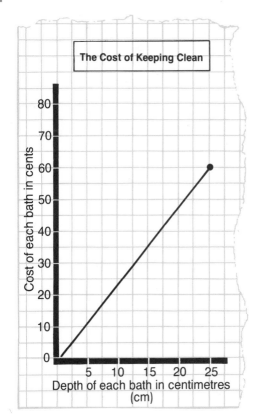

Table 4 shows the recorded measurements of Susan Lee's mass from birth to 18 years. A line graph of these data will show clearly the relationship between Susan's increase in mass and her age.

Table 4 *Age and Mass Data from Susan Lee's Growth Record*

AGE IN YEARS	MASS IN KILOGRAMS (kg)
0 (birth)	3
1	10
2	12
3	14
4	16
5	18
6	20
7	22
8	25
9	29
10	33
11	37
12	42
13	46
14	50
15	53
16	56
17	57
18	57

Drawing a Line Graph

1. Use a ruler to draw an x-axis and a y-axis on a sheet of graph paper (remember to leave margins for labels and scales on each axis).
2. Label the axes.
3. For a graph of Table 4, the scale for the x-axis is easy to choose: Susan's age in years, 0 to 18, at one-year intervals.

   The scale on the y-axis will range from 0 kg to 60 kg. You will have to count the number of squares on your paper and decide what interval to mark on your scale.

4. Now you are ready to plot the points on the graph. Note that the numbers come in pairs: Susan's age and her mass at that age. The graph enables you to show each number pair with just one point. The first point will show Susan's mass at birth (3 kg at 0 years).

   To plot the first point, place the tip of your pencil lightly on the age or x-axis at birth (0 years). Without marking the paper, move the pencil upward along that line on the graph paper until it reaches the 3 kg point on the scale of the y-axis.

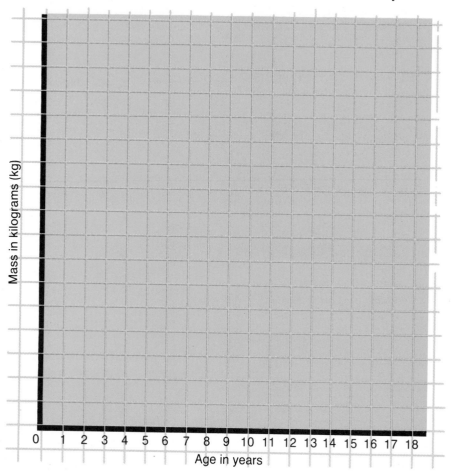

Mass in kilograms (kg)

Age in years

Interpreting a Line Graph

Make a small cross at this position with your pencil.

To plot the second point, place the tip of your pencil lightly on the age axis at 1 year. Again, without marking the paper, move the pencil upward until it comes to the 10 kg point on the y-axis scale. Mark the point as before. Continue plotting the rest of the points in the same way.

5. Using a ruler, draw as smooth a line as you can through the points you have plotted on your graph. Note that the origin is not a point on this graph. Susan's mass was 3 kg at birth, and this should be the first point on your graph.

6. Remember to draw a title block at the top of your graph paper. Give your graph a title and write your name and the date.

7. (a) At about what age did Susan have a mass of 35 kg? (b) At about what age was her mass 26.5 kg?

8. (a) Describe the pattern or shape of your graph in words. (b) Explain why the popular saying "A picture is worth a thousand words" applies to a graph.

9. (a) At what ages did Susan's mass increase most rapidly? (b) When did Susan's mass increase most slowly?

1. Draw a graph of the data in Table 5.
2. Describe the pattern or shape of your graph in words.
3. Line graphs are useful for showing whether there is a relationship between two variables. The graph of the data in Table 5 shows that there is a relationship between distance driven and fuel consumed by the car. Explain how the graph shows this relationship.
4. How is the line or pattern on this graph similar to or different from the graph of Susan's age and mass?
5. Describe the relationship between Susan's age and her mass.

Table 5 *Fuel Used by a Car*

DISTANCE DRIVEN IN KILOMETRES (km)	FUEL IN LITRES (L)
40	2
80	4
120	6
160	8
200	10
240	12
280	14
320	16
360	18
400	20

Divided Circle or Pie Graphs

A pie graph shows the whole of something divided into parts. The parts are shown as wedges of the whole or as "shares" of the "pie." Look at the pie graph here. The circle shows the proportions of various gases in a sample of air.

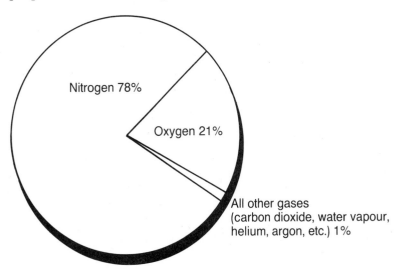

Nitrogen 78%

Oxygen 21%

All other gases (carbon dioxide, water vapour, helium, argon, etc.) 1%

Using a Microscope

Your eyes can only see objects that are larger than about 0.1 mm. Much of the living world is smaller than 0.1 mm—too small to be seen without a **microscope**. A microscope is an instrument used for looking at objects too small for the unaided eye to detect.

The first microscopes used a source of light and one lens to magnify objects. These microscopes are called **simple microscopes**.

The microscopes in the photographs use a series of lenses inside the microscope to improve magnification. They are called **compound light microscopes**, or just **compound microscopes**. The microscope you will use in science class is similar to one of these.

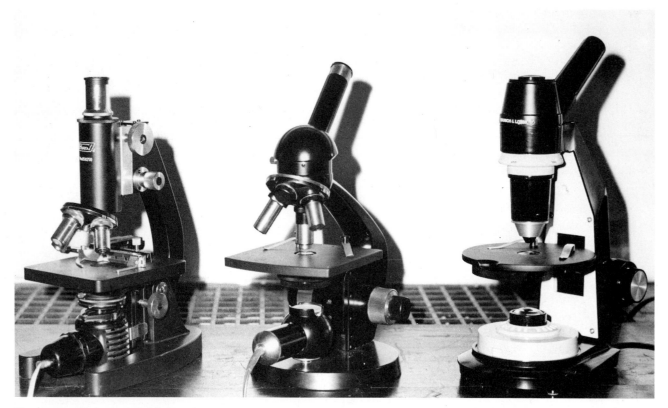

These are all compound light microscopes.

Getting to Know the Microscope

A microscope is a valuable and delicate instrument. Read carefully the hints for handling a microscope. Before you use a microscope you should know how it works. The illustration shows the parts of a compound microscope.

The parts of the compound light microscope.

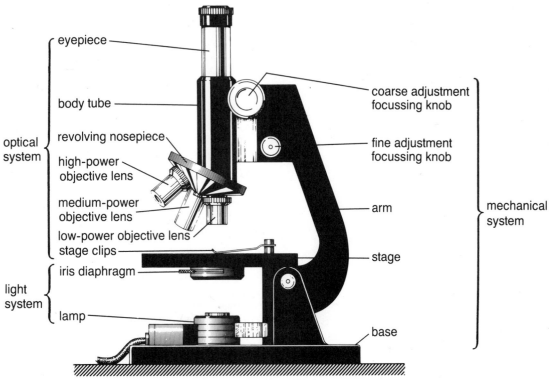

Hints for Handling a Microscope

Read these hints carefully before adjusting and using your microscope. Tell your teacher if your microscope is dirty or if the parts do not move freely. Do not try to force any parts of the microscope to move.

A. Use both hands to carry a microscope. Use one hand to hold it vertically by the metal arm, and use the other hand for extra support under the base.

B. Place the microscope on the surface of a cleared lab bench or a clean desk.

C. Keep the microscope in an upright position when it is not in use and when using liquids. If you do use a liquid, use only a small drop. Keep the stage clean and dry.

D. Start by focusing with the low-power objective lens. (To focus means to make something sharp or clear.) Observe the microscope from the side as you use the coarse adjustment knobs. Lower the low-power lens as close as possible to the stage without touching the stage.

Use the coarse adjustment knobs on either side of the microscope to lower the low-power lens towards the slide. Be careful not to let the lens touch the slide.

Slowly move the coarse adjustment knobs and <u>increase</u> the distance between the lens and the stage to focus the object.

E. When you focus, look through the eyepiece and slowly turn the coarse adjustment knobs so that the lens moves *upward* from the stage. The fine adjustment knobs may then be used to sharpen your view of the object.

> **CAUTION:** Never focus downward.

F. When you have finished using the microscope, remove the glass slide, place both stage clips so that they point forward, and place the low-power lens below the eyepiece.
G. Cover your microscope when it is not in use.

A glass slide prepared for viewing, using water and a square plastic coverslip, is called a **wet mount**. A wet mount keeps objects from being disturbed as you observe them under the microscope.

Skillbuilding Practice 5-1

Preparing a Wet Mount

Materials
letter "a" cut from a newspaper
water
dropper
coverslip
microscope slide
compound microscope

Procedure
1. Using the dropper, place one drop of water on a clean glass slide.
2. Lay the newspaper letter carefully on top of the water.

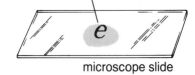

prepared letter on drop of water

microscope slide

3. Take a clean plastic coverslip. Hold it between your thumb and forefinger and touch the edge of the coverslip to the slide and the edge of the water. What happens to the water?

coverslip

4. Lower the coverslip onto the slide so that the letter is covered. (If you have done this correctly, there will be no air bubbles under the coverslip of your wet mount.)

Using the Microscope

Materials

compound microscope
wet mount of a newspaper letter
glass slides and coverslips
facial tissue
photograph from a newspaper or magazine
human hair
piece of paper with a line drawn in ink

other materials suggested by your teacher

Procedure

1. Turn on or adjust the light source of your microscope.
2. Make sure the low-power lens is "clicked" into position.
3. Lay your wet mount of the newspaper letter under the stage clips, with the coverslip facing up and the letter facing you. Make sure the letter is over the opening on the stage.
4. Focus with the low-power lens, using the coarse adjustment knobs.

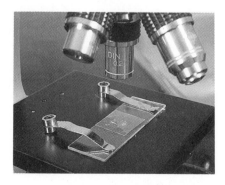

5. In your notebook, draw what you see. Draw a circle around your diagram, and label it to show that you are using the low-power lens.
6. What do you think will happen when you slowly move the glass slide to the right while you are looking through the microscope?
7. Try it and describe what happens.
8. Replace your slide over the opening on the stage. Make any adjustments necessary with the coarse adjustment knob to focus the slide. You are now going to use the other lenses. Follow these steps:
   (a) With the slide in focus, move the revolving nosepiece to bring the medium-power lens into position. Make sure that you hear the "click" that tells you the lens is in place.
   (b) Focus by moving the lens away from the slide, using only the fine adjustment knob.

> **CAUTION:** When using the medium- and high-power lenses, you should use only the fine adjustment knob.

9. Draw in your notebook what you see through the microscope. Draw a circle around your diagram and label it.
10. Using the high-power lens, repeat the procedure.
11. Prepare wet mounts of several other materials. Draw and label a diagram for each material you observe.

Finding Out

1. Describe the difference between the newspaper letter as it appeared to the unaided eye, and as it appeared when viewed under the low-power lens.
2. Explain what happened when you moved the glass slide to the right.
3. Under which lens could you view the whole letter?
4. Suppose you were going to observe the letter "e" under a microscope. Draw your prediction of what you would see under the low-power lens.
5. If you have time, prepare a wet mount and test your prediction.
6. Explain why you must always focus by moving the lens upward (away from the stage) when looking through a microscope.

How Do One-Celled Organisms Move?

Materials

compound microscope
sample of pond water or live
    culture of paramecium
dropper and water
microscope slide
coverslip
tissue paper

Procedure

1. Place a drop of water on a microscope slide.
2. Place a small piece of tissue paper on the drop of water.
3. Put one drop of pond water or culture water on top of the tissue paper. (The tissue prevents the one-celled organisms from moving too quickly.)
4. Complete your wet mount by adding the coverslip.
5. Use low power or medium power to find a paramecium. Observe the outer surface of the cell membrane.

6. Draw and label the structures of the paramecium. Describe the structures that allow the paramecium to swim.

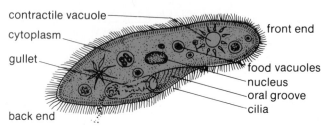

A paramecium in a single-celled organism. It uses its cilia to help it to move.

7. Try to find and draw other organisms that swim in the same manner.
8. Some organisms use a long, hair-like whip called a flagellum to move through a liquid. Try to find organisms, like the euglena, that use a flagellum. Draw them if you can find them.

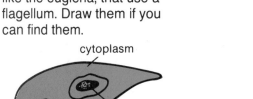

A euglena is a single-celled organism. It uses a flagellum to pull itself through the water.

Glossary

How to Use the Glossary

Use the Pronunciation Key to help you pronounce difficult words.

The numbers in the brackets after each definition tell you where to find the glossary word in the text. The first number is the Unit number or Skillbuilder; the number after the dash is the page number in the text. For example, at the end of the definition of **acid rain**, the number (6-307) tells you that this term is used in the text in Unit 6 on page 307.

Pronunciation Key
a = **a**cid
ah = f**a**ther
e = m**e**t
ih = b**i**te, m**y**
i = th**i**s
oh = h**o**me
uh = tak**e**n, Ann**a**

acid rain rain or snow containing acids formed from gases from cars and some industries combined with rain or moisture in the air. (6-307)

action and reaction every time an object exerts a force (action) on another object, the receiving object exerts an equal force (reaction) in the opposite direction. (3-184)

advancing glacier a glacier where the snout is moving forwards. (6-334)

agar [AY-gar] a jelly-like substance (obtained from seaweed) that is used for growing colonies of micro-organisms. (5-261)

algae [AL-jee] (singular: alga) one kind of single-celled organism. (5-253)

algal [AL-gul] **bloom** a large accumulation of algae that grows rapidly under certain conditions. (5-265)

alloy a metallic substance made from a mixture of two or more different metals. (2-100)

anaerobic no oxygen. (5-275)

annual a plant that lives for only one year or one season. (1-65)

arch a curved structure like a ∩. Arches are often used to support part of a building or bridge. (2-118)

avalanche a fall of mud, snow, ice, and rocks that moves swiftly down a mountain, often growing as it falls. (6-299)

average energy According to the particle theory, some particles in a substance may have more energy than others, but the particles in the substance have an average energy. Temperature is the measure of this average energy of the particles. (4-227)

bacteria (singular: bacterium) one type of micro-organism. (5-253)

balanced forces two forces of equal strength acting on an object in opposite directions. (3-147)

balance scale an instrument for measuring the mass of an object. (3-156)

bar graph a graph showing a series of bars, each of which represents a particular object or quantity. Bar graphs are useful for making comparisons. (SB4-361)

beam a strip of material used horizontally to support a load. (2-102)

bearing a moving part of machinery containing balls or rollers that reduce friction. (3-166)

behaviour the way an organism responds to its environment (1-53)

behavioural adaptation a change in an organism's behaviour in response to a change in its environment (1-53)

bimetallic strip two strips of different metals bonded together; used in some types of thermostats and thermometers. (4-210)

biological weathering weathering that is in part caused by the actions of living things. (6-308)

biologist a scientist who studies living things. (1-53)

botulism a kind of food poisoning caused by a type of bacteria that is common in the soil. (5-275)

boulder a very large piece of rock. (6-316)

brace part of a structure that supports other parts. (2-111)

braided stream a network of shallow streams formed when meltwater from a glacier runs over loose sediment. (6-338)

blubber the fat on whales and other sea mammals. (1-59)

buoyancy The upward push that fluids such as water exert is a buoyant force. (3-175)

buttress a brace that supports an upright structure such as a wall. (2-111)

calibrate to mark a scale of units, such as degrees or newtons, on a measuring device, such as a thermometer or force meter. (3-153)

calving pieces of ice breaking off from a glacier to form icebergs. (6-341)

cantilever a beam or structure that is supported at one end only. A diving board is an example of a cantilever. (2-106)

cartilage a slippery material covering bones in the joints of the human body. Cartilage reduces the effects of friction. (3-168)

cell the basic unit of which all living things (and things that have been living) are made. (1-9)

characteristic a feature. For example, a long neck is the most noticeable characteristic of the giraffe. (1-3)

chemical one of the substances from which all matter in the universe is made. Some examples of chemicals are oxygen, carbon, nitrogen, and hydrogen. (1-10)

chemical additive a substance made by people that is added to food to make it keep longer or look better. (5-284)

chemical energy the energy stored in food and fuel. (4-240)

chemical weathering weathering or breaking down of rocks by chemicals in the air and water. (6-304)

cilia [SIL-ee-uh] hairlike structures of some micro-organisms that wave back and forth so that the organism can move. (1-5) (5-258)

classify to group together things that are alike in some way. You can classify things by colour, by function, by size, or by some other characteristic. (SB2-349)

clay finely weathered particles of certain types of rock. (6-316)

clinical thermometer a thermometer that is used to measure body temperature. (4-214)

cold-blooded animal an animal whose body takes on the temperature of its environment. (1-64)

colony a group of one kind of bacteria living together. (5-256)

column a pillar used to support part of a building. (2-116)

compound light microscope a microscope that uses a series of lenses to increase magnification; the type of microscope used in most schools. (SB5-368)

compression the effect of a load or a force that squeezes or pushes. (2-95)

compressive strength the strength a material has to resist the stress of compression. (2-95)

consumer in biology, an organism that obtains its food by eating other organisms. (5-266)

contaminant a harmful substance that accidentally gets into food. (5-291)

continental glaciers glaciers that cover a large area of land, such as Greenland or Antarctica. (6-333)

contraction the decrease in volume when a substance is cooled. (6-220)

control the part of the experiment in which conditions do not change. The variables in the experiment can be compared with the control. (1-44) (SB-352)

controlled experiment In a controlled experiment, all the variables except one are kept the same. (1-44) (SB2-352)

corrugated Corrugated metal or cardboard is shaped in a row of wave-like ridges. (2-114)

culture a colony of bacteria that are grown by scientists so that they can be studied. (5-261)

cutting a leaf or a piece of stem that is cut from a plant in order to start a new plant. (1-25)

data factual information. (The word "data" is plural; the singular is datum.) (1-10)

decomposer An organism that obtains its food by breaking down wastes and dead materials. (5-266)

delta a triangular piece of land that sometimes forms at the place where a river enters a lake or ocean. (6-324)

deposit to set down. Rock fragments are deposited by moving water or wind. (6-321)

design the way in which the parts of a structure are arranged. (2-76)

diatom [DIE-a-tahm] the most common type of algae, often found on rocks on the bottom of a river. (5-257)

digestion the process of breaking food down into tiny particles that can be used by the body. (1-30)

disperse to spread out or scatter. Seeds from plants may be dispersed by wind, water, or animals. (1-22)

divided circle graph a graph that shows what proportion of a whole are made up of certain items. Divided circle graphs are often called pie graphs. (SB4-361)

dome a large rounded shape, like a series of arches joined at the top. (2-130)

dormant a state in which an organism is alive but not active or growing. Plants or the eggs of an insect are dormant when they are not growing but are waiting for the next warm season. (1-65)

drift rock fragments that have been left behind by a melting glacier. (6-338)

drumlin a tapering hill formed by rocks and earth left behind by a melting glacier. (6-338)

electrostatic force the electricity produced when two surfaces are rubbed against each other. A common word for electrostatic force is "static." (3-171)

end moraine a ridge of rocks and rock particles that are deposited at the tip of a glacier as the snout melts. (6-337)

energy content the amount of energy contained in a substance. Energy content may be measured in joules per gram. (4-237)

environment everything in an organism's surroundings: air, water, all other organisms. (1-27)

erosion the process by which soil and particles of rock are broken up and carried away by water, ice, or wind. (6-310)

esker a long, winding ridge of glacial deposits. (6-338)

evaporation the change of state from liquid to gas. (4-222)

expansion the increase in volume when a substance is heated. (4-200)

extinct A species is extinct if there are no more members of the species alive. For example, dinosaurs are extinct. (1-33)

fermentation a process by which micro-organisms, such as yeast or bacteria, bring about a chemical change in food. Alcohol and carbon dioxide are often produced during fermentation. (5-277)

flagella [fla-JAL-uh] (singular: flagellum) whip-like structures attached to some kinds of bacteria that enable the bacteria to move from place to place. (5-256)

floodplain flat land beside a river that is sometimes flooded by the river. (6-324)

food chain a series of organisms, each of which relies on the organism before it in the chain for its food. For example, in the food chain grass-deer-wolves, deer eat grass and wolves eat deer. (5-268)

food poisoning several kinds of sickness caused by eating food containing certain kinds of micro-organisms. (5-273)

force a push or a pull. (3-142)

freeze-drying a method of drying food in order to preserve it. (5-287)

friction a force that results when the surface of an object moves against the surface of another object. (3-147) (3-162)

function purpose or use. For example, the function of a house is to provide shelter. (2-84)

fungus (plural: fungi [FUNGgih or FUNG-gee]) a type of micro-organism that includes yeast and mushrooms. (5-253)

gas a state of matter. A gas has no definite shape or volume. A gas takes the shape and volume of its container. Steam is a gas. (4-199)

geologist a scientist who studies the Earth. (6-301)

geodesic dome a dome made from a framework of triangles. (2-131)

geothermal energy the energy inside the Earth. We can see this energy in hot springs. (4-243)

gill structure in a fish through which it takes in oxygen and releases unwanted gases. (1-32)

girder a long beam. Girders are often used in buildings and bridges. (2-119)

glaciated valley a valley eroded into a U-shape by the movement of a glacier, often with striated rocks high up the sides of the valley. (6-339)

glacier a large sheet of ice in mountainous regions or near the North and South Poles. (6-332)

gold dust very fine particles of gold. (6-313)

gold nugget a lump of gold. (6-313)

graph a diagram of numerical data that shows changes or comparisons. A graph can make numerical information easier to understand. (SB4-361)

gravel pieces of rock of various sizes, smaller than pebbles and larger than sand. (6-316)

gravity a force that pulls anything with mass toward anything else with mass. (3-156)

groundwater water that sinks into the ground. (6-328)

heat the energy transferred from a hotter substance to a colder one. (4-220)

hibernate to spend the winter in a deep sleep. (1-66)

hoodoo a rock formation consisting of a cap of a hard rock on top of a column of softer rock. (6-302)

horizontal axis the number line of a graph that runs from left to right. The horizontal axis shows the manipulated variable. It is usually called the x-axis. (1-15)

hydroelectricity electricity that is generated from falling or rapidly flowing water. (4-241)

hyphae [HIGH-fee] (singular: hypha) microscopic threads that look like a fuzzy mass of very fine hair. Most fungi consist of hyphae. (5-258)

hypothesis [high-PAW-thuh-sis] a set of ideas or models that provide a possible explanation of why something always occurs in the natural world. (4-220) (SB2-351)

Ice Age a period when ice covered large areas of northern Europe and North America. (6-336)

iceberg a huge chunk of ice that breaks off a glacier when the glacier reaches the sea. The iceberg then floats away. (6-341)

ice wedging a stage in the process of weathering. When water that is trapped in the cracks of a rock freezes, it helps to break the rock apart. (6-303)

inertia the tendency of a stationary object to remain stationary and of a moving object to continue moving unless an unbalanced force acts on it. (3-181)

infer to provide a possible explanation for something observed. (3-165) (SB2-350)

inference a possible explanation for something observed. (3-165) (SB2-350)

infrared light a kind of light that is invisible to the human eye. (4-217)

instinctive the kind of response to a stimulus that an organism makes automatically from birth. (1-57)

insulate to protect something from heat or cold by means of a material that does not transfer heat easily. (2-133)

intervals the divisions between units marked on scales. For example, a scale with marked divisions of 0, 5, 10, 15 has an interval of 5. (SB4-363)

invertebrate an animal without a backbone. (1-18)

irradiation a method of preserving food by using radiation to kill the micro-organisms in the food. (5-287)

joint the place at which two or more pieces of a structure are joined together. (2-122)

joule SI unit for measuring energy. (4-235)

kettle lake a hollow in the surface of the land formed where blocks of ice once melted. (6-339)

landform the naturally occurring surface features of the Earth such as mountains, plains, valleys, rivers, and lakes. (6-296)

lateral moraine see side moraine. (6-337)

learned behaviour that is learned from a model, such as a parent. (1-57)

lichen [LIKE-uhn] tiny living things that grow on rocks. Lichens consist of a fungus and an alga growing together, and depending on each other. (6-308)

life cycle the stages of development that an organism goes through in its life. The life cycle of a human being includes birth, infancy, childhood, and so on. (1-11)

line graph a graph that has a line drawn through the points plotted on it. A line graph shows changes in variables in relation to each other. (SB4-361)

liquid a state of matter. A liquid has a definite volume but no definite shape. Water is a liquid. (4-199)

load the weight carried or supported by a structure. (2-95)

locomotion movement from one place to another. For example, humans use legs for locomotion. (1-34)

lubricant a substance that helps to reduce friction between moving parts in a machine. (3-167)

magnetism a force that acts on objects in the invisible field around a magnet. (3-173)

manipulated variable in an experiment, the manipulated variable is the measurement or condition regulated by the scientist or experimenter. (1-43) (SB2-352)

manufactured something made by people, not found in the natural world. (2-76)

mass the amount of matter in an object. Mass is usually measured in grams and kilograms. An object's mass remains the same whether the object is on Earth or in space. (1-14) (3-156)

material the substance from which something is made. Wood, nylon, and aluminum are examples of materials. (2-93)

matter anything that has mass and occupies space. (3-156)

meander [mee-AN-der] a type of curve in the bed of a river or stream. (6-322)

mechanical weathering weathering that is caused by physical forces, such as water and changes in temperature. (6-303)

melting the change of state from solid to liquid. (4-222)

meltwater water flowing from melting ice, as from a glacier, or from snow. (6-336)

metal fatigue a weakening of metal caused by too much stress. (2-99)

meniscus the curved surface at the bottom of a column of liquid. Because of the meniscus, care must be taken when measuring the volume of a liquid. (SB3-360)

meter an instrument for measuring. (3-148)

microbiologist a biologist who studies micro-organisms. (5-255)

micro-organism a tiny organism. Most are too small to be seen by the naked eye. Many consist of only one cell. (5-253)

microscope an instrument used to look at objects too small for the human eye to see clearly. (SB5-368)

migrate to go to another place to live when the seasons change. Many Canadian birds migrate to warmer countries at the end of the summer. (1-58)

mobile joint a joint that allows part of the structure to move. A door hinge is a mobile joint. (2-124)

moraine a build-up of rocks caused by the movement of a glacier. (6-337)

motion the effect of a force acting on a stationary object. (3-180)

movement any motion or activity that changes the shape, position, or location of an organism. (1-5)

multicellular consisting of many cells. Trees, mice, and humans are multicellular. Some fungi are multicellular. (5-258)

natural anything not made by people is natural. (2-76)

neutral axis an imaginary line along the centre of a beam, halfway between the top and the bottom. At the neutral axis the beam is doing very little to support a load. (2-109)

newton the standard SI unit for measuring a force. (3-153)

non-renewable energy energy from a source that is being used up. Coal, oil, and natural gas are examples of non-renewable energy. (4-245)

nutrient a food an organism eats or absorbs in order to live. (5-267)

optical pyrometer [pie-RAW-met-er] a device that measures temperature by analyzing the light given off by an object as it is heated. (4-216)

organism a living thing. (1-3)

orientation the position in which something is placed. Two possible orientations for a beam may be lying on its wide side or standing on its narrow side. (2-108)

origin the point on a graph where the x-axis and the y-axis meet. (SB4-365)

outwash the deposit left behind by a glacier. (6-338)

ox-bow lake curved lake formed by a meandering stream. (6-323)

particle theory the theory that all matter is made up of tiny particles that are always moving. (4-225)

pebble a small stone or piece of rock. (6-316)

photosynthesis the process of making food using carbon dioxide, water, and energy from the Sun. (1-7)

pie graph see divided circle graph. (SB4-361)

pier a structure that supports a girder or beam. (2-119)

pollination the process by which pollen is carried from one plant to another so that the plants will be fertilized and will be able to produce seeds. (1-25)

predict to tell in advance what might happen on the basis of previous observation and experience. (SB2-350)

producer in biology, an organism that produces its own food. (5-266)

property a characteristic of a material. One of the properties of steel is strength. (2-93)

protozoa a single-celled micro-organism larger than bacteria. (5-253)

range the difference between the largest and smallest of a series of numbers. (SB4-362)

receptor a group of cells in a sense organ that receives messages from the outside world. (1-48)

regeneration the process by which an organism grows new tissue or body parts to replace ones that have been destroyed. (1-18)

renewable energy energy from a source that is not used up when energy is obtained from it. Hydro-electricity is an example of renewable energy. (4-245)

reproduce to produce offspring. (1-6)

resistance the stretchiness or elasticity of a material. A force meter uses resistance to measure the strength of a force. (3-148)

resistance thermometer a thermometer that uses electricity to measure temperature. (4-216)

respond to react to a stimulus. (1-46)

responding variable the data from an experiment that change as a result of changes to the manipulated variable. (1-43) (SB2-352)

retreating glacier a glacier where the snout is moving backwards. (6-334)

rigid joint A joint that does not allow the parts of a structure it joins to move. Rungs are joined to a ladder by rigid joints. (2-122)

runner (plant) a stem of a plant, either above or below ground, from which new plants grow. (1-25)

runoff rainwater that runs across the surface of the ground instead of sinking in. (6-314)

salmonellosis [sa-mon-el-OH-sis or sal-mon-el-OH-sis] a type of food poisoning caused by a species of bacteria. (5-275)

satellite an object in space that moves around another object. The Earth and the other planets are satellites of the Sun. The Moon is a satellite of the Earth. (3-186)

sand finely weathered fragments of rock. (6-316)

scale the equal divisions marked on a measuring device. For example, the divisions marked on a metre stick. (SB1-346)

sediment rock fragments deposited by a stream or river. (6-321)

seed the part of a flowering plant that will develop into a new plant under the right conditions. (1-6)

sense the ability to receive information about the environment. The five senses are sight, smell, hearing, touch, and taste. (1-46)

sense organs the parts of an organism's body that receive information about its environment: eyes, nose, ears, skin, and tongue. (1-46)

side moraine a ridge of rocks that is formed along the side of a glacier as it moves. (6-337)

silt finely weathered particles of rock. (6-316)

simple microscope the earliest kind of microscope using a source of light and one lens to magnify objects. (SB5-368)

soil loose, fine materials from weathered rock mixed with materials from decomposing organisms. (6-308)

snout the tip of a glacier where the ice is melting. (6-334)

solar energy the energy contained in sunlight. (4-244)

solid a state of matter. A solid has a definite volume and shape. Ice is a solid. (4-199)

species (singular and plural) Every living thing is a member of a species. For example, timber wolves, human beings, and Manitoba maples are each members of a different species. (1-3)

spiracle [SPY-ra-kuh or SPI-ra-kuh] tiny holes on an insect's sides that allow the insect to breathe. (1-32)

spore a resting stage in which a bacterium is protected by a thick cell wall. (5-261)

spring scale an instrument for measuring the weight of an object. (3-156)

stalactite [STA-lak-tight] a spike-shaped formation of calcium carbonate hanging down from the roof of a cave. (6-306)

stalagmite [STA-lag-might] a spike-shaped formation of calcium carbonate extending upward from the floor of a cave. (6-306)

standard unit a unit of measurement that always measures the same wherever it is used. (3-153) (SB1-346)

staphylococcal food intoxication a common type of food poisoning caused by a species of bacteria. (5-275)

state of matter There are three states of matter: gas, liquid, and solid. A substance can exist in any of the three states of matter. (4-199)

stationary glacier a glacier where the snout stays in the same position. (6-334)

sterile an object or a substance is sterile if it does not contain any micro-organisms. (5-261)

stimulus (plural: stimuli) anything that causes an organism to react. (1-8) (1-46)

stoma [STOH-mah] (plural: stomata) an opening in the outer skin of a leaf or green stem through which gases can pass. (1-32)

stress a force exerted on an object by a load. Stress may weaken some materials. (2-95)

striation [strih-AY-shun] a scratch in a rock made by rock fragments moved along by a glacier. (6-337)

structural adaptation a special body part that helps an organism to survive in its environment. (1-27)

structure an assembly of parts arranged in a particular way. (2-76)

surface water the water in areas where the water table reaches ground level. Lakes, sloughs, and streams are examples of surface water. (6-330)

technology inventions and other solutions to practical problems. (2-87)

temperature a measurement of the average energy of a substance, or how hot or cold it is. (4-220)

tensile strength the strength a material has to resist the stress of tension. (2-97)

tension the effect of a load or force that pulls. (2-97)

theory a hypothesis that has been supported by experiments. (4-220)

thermocouple a thermometer that uses electricity to measure temperature. (4-216)

thermograph a device that measures temperature by using infrared light. (4-217)

thermostat a device that measures the temperature in a room or an appliance and turns the furnace or appliance on or off to keep the temperature at a desired point. (4-215)

thickened stem a method of plant reproduction by which the stem thickens into a bulb, corm, or tuber, which can produce a new plant. (1-25)

toxin a poison produced by micro-organisms. (5-270)

tributary [TRIB-you-tair-ee] a small stream that flows into a larger stream. (6-315)

unbalanced forces When a force acting on an object is greater than another force acting on the object in the opposite direction, the two forces are said to be unbalanced. (3-147)

valley glacier a glacier in a valley in a mountain region. (6-333)

variation the difference among individual members of a species, such as differences in size, shape, or colour. (1-12)

variable the things that change in an experiment. (1-43)

vertebrate an animal with a backbone. (1-18)

vertical axis the number line on a graph that runs from the bottom to the top. The vertical axis shows the responding variable. It is usually called the y-axis. (1-15)

virus a very small micro-organism that cannot live outside the cell of another living organism. There are many kinds of viruses. (5-253)

vital signs signs of the body's functions that are necessary to life. Vital signs include body temperature, blood pressure, pulse, and breathing rate. (1-41)

volume the amount of space occupied by a solid, liquid, or gas; the amount of space in a container. (4-199)

warm-blooded animal an animal whose body temperature does not change (unless the animal is sick). (1-64)

water table the level of water in the ground. (6-330)

weathering the process by which rocks are broken into small pieces. (6-303)

weight the amount of force that is exerted on an object by gravity. (3-156) (SB5-320)

wet mount a glass slide prepared for viewing through a microscope using water and a plastic coverslip. (SB5-370)

x-axis the number line on a graph that runs horizontally from left to right. The x-axis shows the manipulated variable. It is sometimes called the horizontal axis. (SB4-365)

y-axis the number line on a graph that runs vertically from bottom to top. The y-axis shows the responding variable. It is sometimes called the vertical axis. (SB4-365)

Index